INTRODUCTION TO GOOGLE ANALYTICS

A GUIDE FOR ABSOLUTE BEGINNERS

Todd Kelsey

Apress®

Introduction to Google Analytics: A Guide for Absolute Beginners

Todd Kelsey
Wheaton, Illinois, USA

ISBN-13 (pbk): 978-1-4842-2828-9   ISBN-13 (electronic): 978-1-4842-2829-6
DOI 10.1007/978-1-4842-2829-6

Library of Congress Control Number: 2017945371

Managing Director: Welmoed Spahr
Editorial Director: Todd Green
Acquisitions Editor: Susan McDermott
Development Editor: Laura Berendson
Technical Reviewer: Brandon Lyon
Coordinating Editor: Rita Fernando
Copy Editor: Kezia Endsley
Cover: eStudio Calamar

Distributed to the book trade worldwide by Springer Science+Business Media New York, 233 Spring Street, 6th Floor, New York, NY 10013. Phone 1-800-SPRINGER, fax (201) 348-4505, e-mail orders-ny@springer-sbm.com, or visit www.springeronline.com. Apress Media, LLC is a California LLC and the sole member (owner) is Springer Science + Business Media Finance Inc (SSBM Finance Inc). SSBM Finance Inc is a **Delaware** corporation.

For information on translations, please e-mail rights@apress.com, or visit http://www.apress.com/rights-permissions.

Apress titles may be purchased in bulk for academic, corporate, or promotional use. eBook versions and licenses are also available for most titles. For more information, reference our Print and eBook Bulk Sales web page at http://www.apress.com/bulk-sales.

Any source code or other supplementary material referenced by the author in this book is available to readers on GitHub via the book's product page, located at www.apress.com/9781484228289. For more detailed information, please visit http://www.apress.com/source-code.

Printed on acid-free paper

Apress Business: The Unbiased Source of Business Information

Apress business books provide essential information and practical advice, each written for practitioners by recognized experts. Busy managers and professionals in all areas of the business world—and at all levels of technical sophistication—look to our books for the actionable ideas and tools they need to solve problems, update and enhance their professional skills, make their work lives easier, and capitalize on opportunity.

Whatever the topic on the business spectrum—entrepreneurship, finance, sales, marketing, management, regulation, information technology, among others—Apress has been praised for providing the objective information and unbiased advice you need to excel in your daily work life. Our authors have no axes to grind; they understand they have one job only—to deliver up-to-date, accurate information simply, concisely, and with deep insight that addresses the real needs of our readers.

It is increasingly hard to find information—whether in the news media, on the Internet, and now all too often in books—that is even-handed and has your best interests at heart. We therefore hope that you enjoy this book, which has been carefully crafted to meet our standards of quality and unbiased coverage.

We are always interested in your feedback or ideas for new titles. Perhaps you'd even like to write a book yourself. Whatever the case, reach out to us at editorial@apress.com and an editor will respond swiftly. Incidentally, at the back of this book, you will find a list of useful related titles. Please visit us at www.apress.com to sign up for newsletters and discounts on future purchases.

—*The Apress Business Team*

Contents

About the Author

Todd Kelsey, PhD, is an author and educator whose publishing credits include several books for helping people learn more about technology. He has appeared on television as a featured expert and has worked with a wide variety of corporations and non-profit organizations. He is currently an Assistant Professor of Marketing at Benedictine University in Lisle, IL (www.ben.edu).

Here's a picture of one of the things I like to do when I'm not doing digital marketing—grow sunflowers! (And measure them. Now there's some analytics for you!)

I've worked professionally in digital marketing for some time now, and I've also authored books on related topics. You're welcome to look me up on LinkedIn, and you're also welcome to invite me to connect: http://linkedin.com/in/tekelsey

About the Technical Reviewer

Brandon Lyon is an expert in SEO, SEM, and Social Media and Web analytics, and is President of Eagle Digital Marketing (https://www.eagledigitalmarketing.com), a full-service agency in the Chicago area. When he isn't advising local business owners and CEOs of mid-sized companies, he enjoys hockey and doing his best to survive the occasional subzero temperatures. Brandon enjoys helping companies face the challenges of the future with optimism, including navigating the treacherous waters of the Amazon ecommerce river, and taking advantage of the goldmine in marketing automation.

Introduction

Welcome to web analytics!

The purpose of this book is to provide a simple, focused introduction to web analytics, and Google Analytics specifically. It's geared toward employees who may be working at a company or non-profit organization, for students at a university, or for self-paced learners. The approach is the same one that I've taken in most of my books, which is conversational, friendly, with an attempt at making things fun.

The experiment is to find a way to help people get started with digital marketing in a way that is fun and helps build skills—maybe through an internship, paid work, volunteer work, freelance work, or any other type of work. The focus is on skills and approaches that can be immediately useful to a business or non-profit organization. I'm not going to try to cover everything, but just the things that I think are the most helpful.

The other goal is to help you leave any intimidation you have in the dust. I used to be intimidated by marketing, and now look at me. I'm a marketing strategist and an assistant professor of marketing! But I remember the feeling of intimidation, so part of my approach is to encourage any reader who may feel uncertain about the field.

The fact is that web analytics has a lot of options, and there's a lot of material out there. It can be overwhelming! But it can also be very doable, if you leave intimidation in the dust, take incremental steps, try things out, and build your confidence.

For example, I had a friend who used to be a journalist, and he was looking for new career opportunities. I helped get him started in digital marketing, and one of the first things he ran into was feeling overwhelmed by all the options, including all the articles about all the options. "There are so many tools out there!," he used to say, "How am I ever going to learn all of them?!?"

The answer is that you don't need to learn all of them. No one can. The thing to do is focus on trying some of the tools and skills and go on from there. I encouraged my friend not to worry about trying to learn everything, but instead to just learn some basics.

The friend worked with the basics, gained experience, and was able to find a local agency that gave him a shot at doing some freelance work. The career didn't easily develop for him—he had to put effort into it. But a few years later, he's doing full-time freelance work in digital marketing. He was able to leave intimidation in the dust, and I believe he's also had some fun with it too.

LinkedIn showed digital/online marketing as a top skill that got people hired in 2013/2014, and web analytics is one of the core skills for online marketing—this includes being able to understand the performance of web sites and ad campaigns. Analytics is considered a part of *business intelligence*, which also figures prominently on the list:

Each year the way they refer to digital marketing seems to change, but since 2013, digital marketing (of which analytics is a core part) has been at the top. Demand will fluctuate over time, but we are talking about the top skills in any field that get people hired.

- 2014: https://blog.linkedin.com/2014/12/17/ the-25-hottest-skills-that-got-people-hired- in-2014

- 2015: https://blog.linkedin.com/2016/01/12/ the-25-skills-that-can-get-you-hired-in-2016

- 2016: https://blog.linkedin.com/2016/10/20/ top-skills-2016-week-of-learning-linkedin

One of the other things I've seen in my career, which I try to reinforce in these books and in my classes, is the way that the core areas of digital marketing are related. For example, I consider web analytics to be tightly connected to all other areas in digital marketing. Content is key in digital marketing, as you'll learn if you read my book *Introduction to Search Engine Optimization* (Apress, 2017), and advertising campaigns on search engines (AdWords) and social media take content and put it out there.

At the end of the day, you need analytics to measure the performance. That makes it super important.

This book mentions what I call the core areas of digital marketing: Content, AdWords, Social, and Analytics (CASA for short). My goal is to reinforce how all the areas are connected. AdWords is Google's tool for creating ads for search engine marketing. The inspiration came from my professional background, as well as looking at trends in the marketplace.

The Core Areas of Digital Marketing

C Content/SEO: search engine optimization is the process of attempting to boost your rank on Google so that you get higher up in search rankings when people type in particular keywords. Higher in search rankings = more clicks. The top way to boost rank is to add quality content that is relevant for your audience.

A Adwords: the process of creating and managing ads on Google (Adwords), where you attempt to get people to click on your ads when they type particular keywords in Google. You pay when someone clicks.

S Social Media Marketing: the process of creating and managing a presence on social media, including making posts, as well as creating advertisements. The main platforms are Facebook, Twitter, and YouTube, as well as Instagram and Pinterest

A Analytics (Web visitors): You can gain valuable insights when you measure the performance of your websites and advertising campaigns. Google Analytics allows you to see how many people visit your site, where they come from and what they do.

Best wishes in learning web analytics and Google Analytics!

Overview

This chapter takes a look at what *analytics* is and introduces related concepts. This chapter, and the entire book, is oriented toward beginners. My goal is to encourage you to consider learning more about analytics, including using a tool called Google Analytics, and to see if I convince you that it actually is fun.

Web analytics is becoming increasingly important to online marketers, as they seek to track return on investment (ROI) and optimize their web sites. We'll learn about Google Analytics, starting with creating a blog and monitoring the number of people who see the blog posts and determining where they come from.

Don't Be Alarmed: Analytics Can Be Fun

So I remember when I started working in various jobs after college, that one thing I knew for certain is that I didn't feel like I was a "numbers" person. Accounting, finance, or any other type of numbers always seemed foreboding, and outside my experience and comfort zone. It was the last possible door I wanted to walk through, and to get me through it, you'd have to drag me.

© Todd Kelsey 2017
T. Kelsey, *Introduction to Google Analytics*, DOI 10.1007/978-1-4842-2829-6_1

But the interesting thing was that I got my first taste of analytics without even realizing it.

I created a web site with a friend, and we wanted to know how many people were visiting it and where they were coming from, so we searched for a tool that would help us with that. That tool became a part of our toolbox.

It was *fun*. Getting to see who was coming, what the traffic was like, was really interesting.

Then, in later work experiences, including some where there had been transition in the companies I worked for, including layoffs, I became more sensitive to how these roles related to the overall business.

At one point, a mentor gave me some advice that really helped me. She was experienced and she said to me, "Todd, you need to follow the money trail".

She wasn't saying, "bow down to money." She was just saying, it's good to understand how money flows through a business—what makes money and what costs a business money.

My perspective on money and finance was challenged, and I realized that it would probably be a good idea to consider not just what I felt like doing, but what would be a benefit to a business or organization—especially during hard times, such as a recession, or competition, etc. I also started to understand that the kinds of skills and roles that had a direct impact on helping a company succeed were in high demand.

This path led me to pay more attention to online marketing in particular, as well as social networking to a particular degree. In the midst of hard economic times, Google was increasing in value, at a time when many or even most companies were having serious financial issues. It was partly because Google was helping companies do a good job of tracking ROI with Google ads—AdWords. AdWords helps businesses know what they are making based on what they are spending in terms of online ads.

As I grew and matured, I also realized that web analytics was an important skill and I started learning more about it. It helped me find work and be competitive.

While I'm still not "passionate" about numbers, I do see things like Google Analytics as an important tool. Maybe I'm more in touch with my inner analyst.

My recommendation is, seriously consider learning web analytics, in order to strengthen your career. It won't hurt, and it can also be fun.

Another thing I suggest with web analytics in particular, and in any situation where you are dealing with numbers based on purchases, is to remember this aphorism. It may look like a bunch of numbers:

But it's really not about the numbers, in the end. It's about the people:

designed by ⚡ freepik.com

Personal ROI: How Analytics Will Help Your Career and Your Organization

Learning analytics can have a significant impact on your career, regardless of the area of digital marketing you work in.

Social Media Marketing

It's increasingly important for social media marketers to understand how to measure and optimize the performance of campaigns—you might call this "social analytics". If you haven't read my *Introduction to Social Media* book yet, you might want to have a look. It includes some coverage of social analytics. You could also work your way through this book and then head on over to that one, to see the connections.

Digital/Online Marketing

The goal of digital marketing is often to sell something, or at least to get people to visit and sign up for something. So web analytics is a crucial tool for monitoring how your efforts are going.

Business Intelligence

You could think of this as "advanced analytics". Business intelligence might include web analytics (reviewing the performance of your web site and associated marketing), but it can also extend into other areas, such as "competitive intelligence," by using a tool like compete.com or just looking at financial trends. In my own experience, starting out with web analytics helped me understand how online marketing and reports fit into overall business intelligence. It would be fair to say that business intelligence is ROI.

You don't need any skills to start off, and you *don't* need to be a numbers person. This book is for anyone who wants to get a job in online marketing or who wants to learn how the performance of their web sites fits into the business model. Google Analytics is one of the top tools, and web analytics can be a competitive differentiator in the job market, whether it's one part of a skillset or a dedicated role.

Note This book is intended to help students to view web analytics info and learn how to develop insights. Skills in this area connect to other areas, such as search engine marketing and social media marketing, and provide a network effect to help students become more effective online marketers, as well as more employable.

Here's a suggestion and invitation, for your "personal" ROI, that I want you to consider—become Google Analytics qualified. This book is an introduction to the concepts and the tool, and then I'll point you to more learning material that Google has, which you might want to pursue in order to get qualified. See Chapter 8 for more details.

Basically, having this qualification on your resume or LinkedIn profile will help to show your credibility, to your colleagues and potential employers. In short, it will help your career.

Some certifications cost thousands of dollars to prepare for, and the tests can be expensive too—however, at the time of writing, Google qualification is free.

Organizational ROI

I don't know if "organizational return on investment" is really a term, but I guess now it is.

What I mean is that even if you already have a job somewhere, going for Google qualification can help any company or organization you are a part of. Not only will it help with your credibility when people interact with you on sites like LinkedIn, but it will also help you think more about tracking ROI. It's a mindset that will be a benefit to any business, to help you make good choices about which direction to go in.

Free and Corporate Tools

There are a variety of web analytics tools out there, which people use on a free and paid basis. For grins, try going on LinkedIn and doing a job search for "web analytics" to see what comes up. Chances are that one or more of these programs will be mentioned.

Google Analytics

Google Analytics is free, and it is always increasing in power, to rival and in some cases exceed the performance of the "paid" tools. For example, you'll often see companies using a combination of Google Analytics and paid tools. Google Analytics can't do everything, but it's a good place to start.

Adobe Analytics: Omniture

Adobe Analytics/Omniture has long been considered one of the top "enterprise" web analytics tools. It has a lot of power, sophistication, and customization. There are things that Omniture can do that Google Analytics can't, and vice versa. Knowledge of this program can definitely help you get a job or a higher salary. It's a somewhat chicken and egg situation. It's an expensive program, with no trial version at the time of this writing, so few people know how to use it well. Hence, it is harder to find people with this skill; therefore, the demand (and salary) can be higher.

If you are going into online marketing and you can find a company to intern at or work with where you have the opportunity to learn Omniture/Adobe Analytics, that could be a good opportunity.

More recently, larger companies have been consolidating smaller companies and developing integrated "marketing clouds," so you should look at some of the tools that Adobe Analytics is connected with. See www.adobe.com/analytics.

Other "enterprise" options include tools like WebTrends. Hopefully at some point enterprise folks will wise up and offer trial versions to help people get access and learn to use them.

Open Source Analytics

For people who are interested in completely controlling their own data, open source analytics programs may be an option. In certain cases, they may allow you to have the functionality you need, without giving up any of the value. For example, most people and companies accept the value proposition when using a Gmail address, doing a Google search, or using Google Analytics. They know that Google will analyze the data and make money off of it. In other words, when you search for something, Google might display an ad based on your behavior, which you might be interested in.

Google isn't in the business of selling your contact information, per se, but with analytics, it might gather your information, make it anonymous, and group it with a lot of other data, with the goal of making money off of it. Google can also somehow use related "cookie" information.

Google isn't doing anything illegal, and personally, I don't think there's anything to worry about. But you might be interested to just take a peek at some of the open source options out there, in case you end up being more concerned about the data someday, or you end up with a client or employer who is. There are some more value propositions on their sites.

See http://piwik.org and http://www.openwebanalytics.com/.

As for me, and many thousands of businesses, I'm cool with Google Analytics.

Social Analytics

Social analytics is an area in which particular social media channels, such as Facebook, YouTube, etc., allow you to get information about how your social media efforts are working. You can determine the number of people who like your page, follow you, or talk about you— that kind of thing. There aren't any dominating "all in one" social analytics tools as of yet, but take a look at Adobe's "social analytics" offering, to see an example of an attempt to become one.

In general, the social analytics tools are free and built into social media. In some cases, there are low cost tools like Hootsuite (which also has a free version) that will draw some of the material together for you.

If any of that sounds interesting, take a look at my book entitled *Introduction to Social Media Marketing* (Apress, 2017).

My general advice for beginners is to have some fun learning Google Analytics and also try a bit of social media marketing. Get your feet wet with social analytics.

Things Change

As you explore these areas, be prepared for things to change, but don't worry. You don't need to learn every tool. I recommend taking an incremental, gradual approach.

As with social media marketing, there are a lot of options out there, and if your eyes start to glaze over with the mention of all these analytics tools, don't be alarmed! Don't try to learn them abstractly—try them out. I'll introduce them to you in a way that is fun.

Search Drill

If you're just getting acquainted, I suggest doing a few Google searches to see what's out there and reading anything that looks interesting:

- YouTube Google Analytics intro
- Learning Google Analytics
- Understanding web analytics YouTube
- What the heck is Adobe analytics

Learning More

If you want to click on a link and learn something, try these videos:

- Web Analytics—The Basics: https://www.youtube.com/watch?v=1lfnOYuOzxA
- Successful Web Analytics: https://www.youtube.com/watch?v=bpDxGrSqA-E

Conclusion

Congratulations on making it through this chapter! I'm trying to make the approach in this book as friendly and relevant as possible, based on the intimidation I used to feel. Feel free to let me know if it's working, if you feel like you're still intimidated, or if you feel like it's "too fluffy". Best wishes in learning analytics!

Blogalytics

This chapter is a basic recipe to get started in analytics, to take the first step in being able to track web traffic and see what's happening. You'll create a simple blog, start a Google Analytics account, connect the two, and discuss things along the way. The goal is to make a simple, relevant way to start exploring analytics.

Create a Google Account/Gmail Address

Google has a lot of free tools that make it easier to work with content, and when you have a Google account, it just makes it easier to sign in to all the tools.

As a first step, I recommend creating a free Google account by going to http://mail.google.com and clicking Create Account.

Start a Blog

To get started, visit http://www.blogger.com and either sign in with your Google account or click the Create an Account link at the bottom of the page.

© Todd Kelsey 2017
T. Kelsey, *Introduction to Google Analytics*, DOI 10.1007/978-1-4842-2829-6_2

Then on the blogger site, click the New Blog button.

For practice, I wouldn't be too concerned with the title. You can change it later easily, and you can also create/delete blogs easily. Feel free to try "Social Media Perspective" as a title.

Blogs List › Create a new blog

Title   Social Media Perspective

Address   .blogspot.com

You can add a custom domain later.

The title is simply what appears visually at the top of the blog. The Address is the opportunity Google gives you to have a custom address. Because it's a free tool, you might have to experiment a bit until you find one that's available. Type in ideas in the Address field and see what happens:

socialbuzz| .blogspot.com

Sorry, this blog address is not available.

What you're doing is coming up with the custom portion of the blog address.

socialbuzznews| .blogspot.com

This blog address is available.

It turns out for our example, the address socialbuzznews.blogspot. com is available. The link for this blog would be http://socialbuzznews. blogspot.com.

Tip You'll want to make a note of this link to your blog, so you can use it when you start a Google Analytics account later in this chapter.

After you choose a title and address, you can choose a template for the look and feel of the blog. You can also change the template later:

Blogs List › Create a new blog

Title	Social Media Perspective
Address	socialbuzznews| .blogspot.com ✓
	This blog address is available.

Template

Simple	Dynamic Views	Picture Window
Awesome Inc.	Watermark	Ethereal

You can browse many more templates and customize your blog later.

Create blog!   Cancel

After you select one (I recommend starting with Simple), click the Create Blog! button.

With these simple steps, you've created a blog and can start blogging!

Your mission if you should choose to accept it is to make a sample post, and then share the link on Facebook or via e-mail with someone.

Note One way to "cheat" if you forget the link for your blog is to click on the View blog button (see the screenshot above), which will open the blog in your browser. Then you can copy the link from the Address field and paste it into Facebook or into an e-mail, etc.

To learn more about Blogger, access the settings menu (the little gear icon) when you're signed int o Blogger and select Blogger Help:

There are a variety of helpful articles:

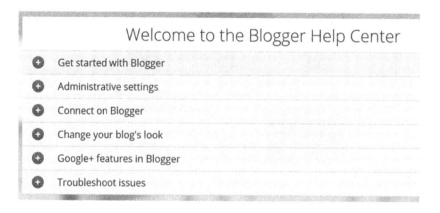

You can always go directly to help with this link: https://support.google.com/blogger.

Start a Google Analytics Account

The next step is to create a Google Analytics account, which you do by going to `http://www.google.com/analytics/`.

If you're already signed into Gmail/Google, you'll see an Access Google Analytics button:

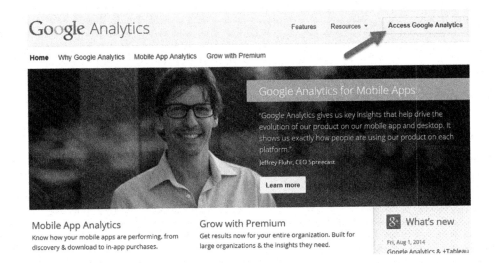

Otherwise, you can click Sign In and sign in with your Gmail/Google account, or click Create an Account:

Sign in or create an account

That's the sign-in process. Once you're signed in as a Google user, you should get a page like this one, where you can click the Sign Up button to create your Google Analytics account:

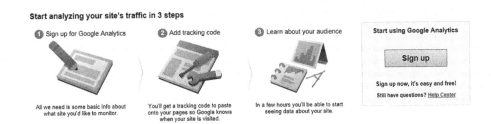

Google Analytics defaults to creating an account around web sites, but if you're interested in mobile marketing, at some point you might like to explore the idea of developing an app and integrating Google Analytics directly in the app.

To begin, type an account name—it could be your name, the name of your organization, or something like Learning Analytics. It really doesn't matter at this point:

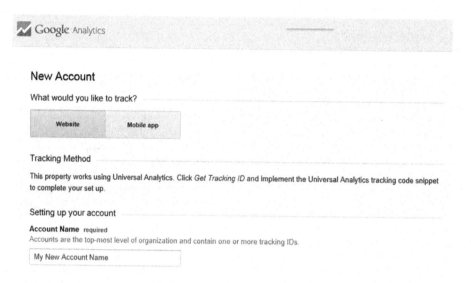

Next, you set up a *property*. Whenever you add a new site or blog for Google Analytics to track, it's considered a property. This is where you paste in your blog's address, if you're following along:

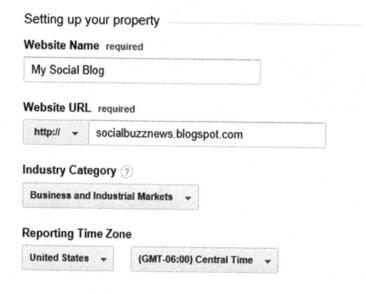

Then click Get Tracking ID:

Tracking IDs are basically codes that Google gives you to place on your site or blog, so that the site can "talk" to Google. The ID is unique for every site/blog you set up. It's a one-time, thing, and you end up taking it back to your site, in order to set up the connection.

Next, click the I Accept button at the bottom of the agreement. You might need to roll your mouse over the top of this window until it becomes four arrows, and then click and drag it up, in order to see the I Accept button.

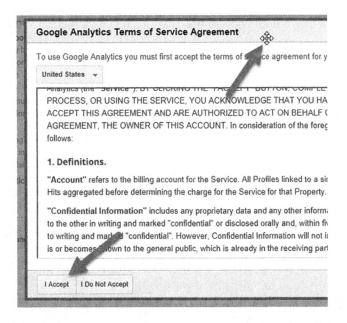

Then, Google Analytics should open with a page, based on the name you gave to your site or blog:

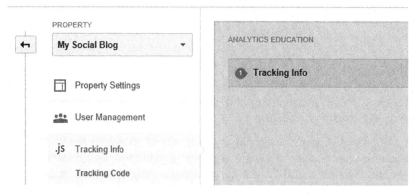

You can scroll down until you see the Tracking ID (it's UA-53515736-1 shown here):

Tracking ID
UA-53515736-1
Website tracking

This is the Universal Analytics tracking code for this property. **To get all the benefits of Universal Analytics for this property, copy and paste this code into every webpage you want to track.**

Then just select it, right-click (Windows) or Ctrl+click (Mac), and choose copy:

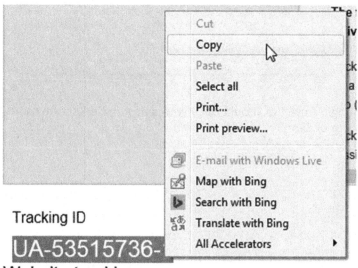

Even if you copy this code into memory (for pasting into Blogger), I recommend making a note of it somewhere.

Connect Google Analytics to Blogger

Now that you've created a Google Analytics account and generated a tracking code, you can go back to your blog and click Settings:

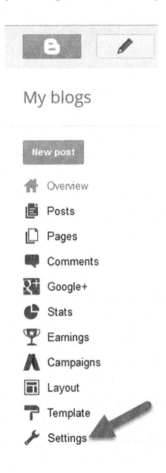

Under Settings, click Other:

Settings

Basic

Posts and comments

Mobile and email

Language and formatting

Search preferences

Other

Find the section that says Google Analytics:

Google Analytics

Analytics Web Property ID   ?

Paste that marvelous tracking code you copied before (from *your* site or blog):

Tracking ID

UA-53515736-1

Website tracking

Into the Analytics Web Property ID field:

Google Analytics

Analytics Web Property ID   ?   UA-53515736-1

Then click Save Settings:

Save settings

Looking at Google Analytics

In theory, Google Analytics should now be connected. It can take 24 hours for anything to happen, but in the meantime, I suggest making a post on your blog and sharing it with friends on Facebook as a test.

All you really need at this point is a few clicks on your blog, just to test things out a bit.

In order to see how things are going, go to `http://www.google.com/analytics`, sign in, and click on Home:

You should see a section on your Google Analytics page. It's based on how you named your Google Analytics account, including the name of your site, and finally a little globe icon with your site:

To see how things are going, click on your "property":

In theory, you should see something like this image, which is the kind of thing you'll be learning about. This is web analytics:

At a very basic level, what programs like Google Analytics seek to give you is a sense of your web site's performance—that is, how many people visit, and their behavior on the site.

Using a simple example like this is a good place to start, and ultimately what most web analytics tools help you do is track revenue or goals, such as people visiting and then purchasing or signing up for something.

To scratch the surface, after you generate data over time, one feature to keep in mind about analytics programs is the way you set a date range.

In Google Analytics, you can click on the arrow to the right of the "default" date:

Then you can choose from several options:

You can regularly look at Google Analytics and review how things are going on a consistent basis. With analytics, you are often looking back and looking for trends. Are visits increasing? Did a particular campaign increase traffic? Things like this.

When you set a new date range, you need to click Apply:

Apply

Don't be afraid to review other sections in Google Analytics on the left side. In some cases, you need to be gathering data for a while for these sections to generate meaningful information. It's interesting at this point to note the kinds of things you can track:

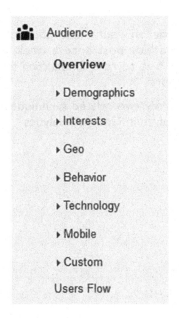

Learning More

If you're eager to learn more about Google Analytics, the Help Center has some good introductory material at https://support.google.com/analytics/?hl=en#topic=3544906.

These resources may also be of interest:

- How to Install Google Analytics on Blogger: http://www.lacquerheadsofoz.com/2013/04/how-to-install-google-analytics-on.html

- How to Add Google Analytics to Blogger 2014: https://www.youtube.com/watch?v=DOF7jCovhpQ

Conclusion

Congratulations on setting things up!

I think the reason analytics can be fun is that you can see "under the hood" of a web site and look at how many people are visiting. If you've used Facebook, it's similar to seeing how many people like your post or page. But analytics takes it from a single item to a fuller picture of what's going on.

In order to have fun, the next step is to get some traffic to your blog or site. One thing you can do is write a blog post and share it wherever you can think of to get some clicks, including Facebook.

Consider regularly blogging on your exploration of online marketing and analytics, such as making a blog post once a week, to see what it's like to get traffic over time. Feel free to read *Introduction to Social Media Marketing* (Apress, 2017) to learn more.

In the next chapter, you'll try two related techniques to "jump start" things and get some traffic to analyze in Google Analytics.

Getting Traffic for Analytics

This chapter looks at a couple of basic techniques for generating traffic on a blog or web site, following the previous chapter about creating a blog. The chapters are written as self-contained experiments. Ultimately, you need traffic in order to be able to look at analytics, so this chapter describes a few ways to generate traffic.

Basic Social Media Promotion

One of the easiest ways to get some test traffic to your blog or web site is by sharing a link to a post on any social media accounts you have. (If you don't have any, I suggest signing up for Facebook, Twitter, and LinkedIn).

If you want more connections, you can always use the feature where you share your e-mail contact list, and the programs suggest new connections. Check out these sites for good advice:

- How do I import friends from other accounts?: https://www.facebook.com/help/189976267783273

- "How to import contacts into Twitter: http://www.dummies.com/how-to/content/how-to-import-contact-lists-to-twitter.html

© Todd Kelsey 2017
T. Kelsey, *Introduction to Google Analytics*, DOI 10.1007/978-1-4842-2829-6_3

- "How to import contacts into Linkedin: https://www.linkedin.com/help/linkedin/answer/4214?query=import%20contacts%20into%20linkedin

Once you have some connections to work with, the next step is to take the link for your blog and post an invitation to social media. This isn't about social media promotion per se—this is just an exercise to generate traffic, for the purpose of being able to look at your analytics data.

Get the Link

Go to the blog you created in Chapter 2 and copy the link into memory.

For this blog, the "general" link is http://npoex.blogspot.com/.

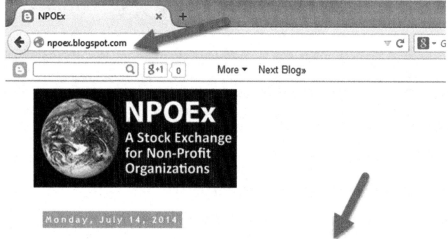

There's a difference between the blog link and a link to individual blog posts. Click on the title of a blog post you made, and you'll see a longer link. This is the one you'll want to share on social media:

http://npoex.blogspot.com/2014/07/non-profit-stock-exchange-user.html

Tip Visuals are good. Include a picture in the blog post. When you post to social, to get more notice, include a picture in your post. Technically you could post a photo to Facebook, and include a link in the post, but what you really want to do is include visuals in your original blog post. Then when you post to social media, in some cases (Facebook/LinkedIn), it gives you the option of having a visual preview. This will garner more attention, more clicks, and ultimately more data to play with.

Make your blog posts ongoing. For learning analytics, I recommend that you get in the habit of posting once a week to a blog (or web site) that is connected to Google Analytics, and then look back at it. To generate ongoing data, add the Follow by Email feature in Blogger. When people visit, they can sign up to get posts automatically when you post them. In theory, this should help you get some ongoing clicks.

Subscribe for updates by Email

| Email address... | Submit |

Make Your Pitch

This chapter isn't meant to replace my *Introduction to Social Media Marketing* book. I recommend you read that too, because it's all connected. But we're just taking a simple, quick tour of how to get some clicks.

So when you post to social, make a pitch for clicks. Tell people what you're up to and invite them to click on the link below. Something like this:

> *Friends and colleagues, please click on this link. I'm doing a test for a book I'm writing to help people learn web analytics, in order to show readers how they can share their blog posts on social media. Every click will help generate some "data" that I can look at in the book. You're also welcome to subscribe by e-mail if you're interested in the topic. (Please feel free to pass this along.)*

```
http://npoex.blogspot.com/2014/07/non-profit-stock-
exchange-user.html
```

Post It to Social Media

Put your post on social media. Notice how Facebook picked up the image from the blog post in the preview.

You may not realize that you can do the same thing on LinkedIn.

When you log in, there should be a spot to share an update:

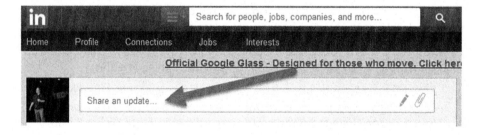

It's worth trying.

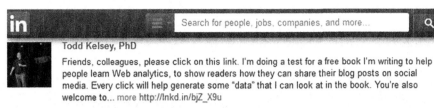

Todd Kelsey, PhD

Friends, colleagues, please click on this link. I'm doing a test for a free book I'm writing to help people learn Web analytics, to show readers how they can share their blog posts on social media. Every click will help generate some "data" that I can look at in the book. You're also welcome to... more http://lnkd.in/bjZ_X9u

Non-Profit Stock Exchange: User Experience - 1

npoex.blogspot.com · First of several "User Experience" (UX) brainstorms. Today I'm going to try dipping my toes lightly but concretely into what it might be like to try a stock exchange for non-profits. A technical term for this is called...

Like · Comment · Share · 1s ago

Promote Your Post

I get kind of an icky feeling even sharing this technique, but it's built into Facebook, and it feels a bit like a mafia shakedown. See *Introduction to Social Media Marketing* (Apress, 2017) for my skeptical perspective on how advertising on Facebook has evolved.

Whether it is because there are too many posts by too many people, or whether Facebook has intentionally limited who sees your posts, the bottom line is that not as many people will see your posts on your personal profile as in the past, unless you pay Facebook to promote them.

The quickest, easiest way to get your posts noticed is to click the Promote link at the bottom of the post and pay the Facebook mafia:

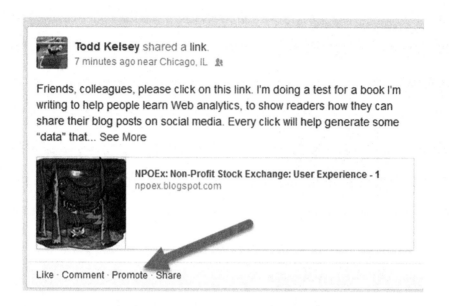

Todd Kelsey shared a link.
7 minutes ago near Chicago, IL

Friends, colleagues, please click on this link. I'm doing a test for a book I'm writing to help people learn Web analytics, to show readers how they can share their blog posts on social media. Every click will help generate some "data" that... See More

NPOEx: Non-Profit Stock Exchange: User Experience - 1
npoex.blogspot.com

Like · Comment · Promote · Share

This extortion is probably worth the money, at least as a learning experience:

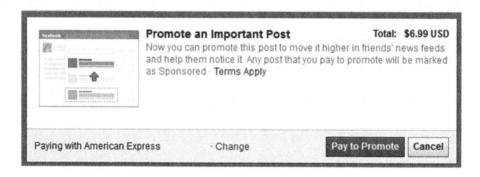

Promote an Important Post   Total: **$6.99** USD
Now you can promote this post to move it higher in friends' news feeds and help them notice it. Any post that you pay to promote will be marked as Sponsored · Terms Apply

Paying with American Express   · Change   Pay to Promote   Cancel

Enter your credit card information and click Pay to Promote:

It's a one-time, limited thing. A bit like blackmail. Blackmail payments have a way of growing, and not going away. So too with social networks' interest in monetizing everything—even if you had the impression that it shouldn't cost anything to share with the network of friends you built. I suppose Facebook provided the platform to build that network, but still. But for grins, lets give Facebook the benefit of the doubt. Personal social advertising is just like professional social advertising. You pay to get exposure.

Wait, isn't social media supposed to be "free" advertising?

At any rate, when you pay Facebook to promote your post, it shows the word "Sponsored" at the bottom:

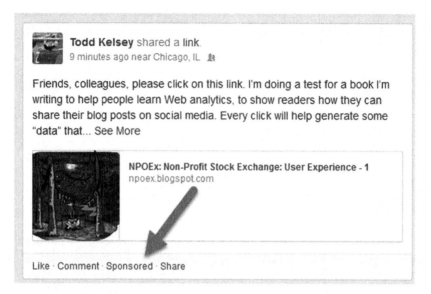

::Shudders::

Make an Ad

Another way I suggest trying to add more clicks is a simple Facebook ad campaign. As much as Facebook ads have changed as Facebook has grown, it's still probably easier to create a Facebook ad than a Google ad, for example. Although I should probably take a poll on that, from people who have tried both.

So we're just going to take the simplest way possible through the process, without getting into specifics. It's all about the data.

Go to https://www.facebook.com/ads/create and click on the Clicks to Website option:

What kind of results do you want for your ads?

Then, copy your trusty blog link, the one to the specific post, and paste it in the field:

Until/unless Facebook changes, when you paste the link, it may flip you to another screen. Don't be alarmed.

On the next screen, you'll want to upload an image. It recommends a large image, and if you try to run the ad with a smaller image (like the one show here that Facebook picked up from the blog post), it won't run the ad. If you're just testing things, search for "Earth" on Google. Click the image link at the top of the Google page, right-click (Windows) or Ctrl+click (Mac), and then save the image to disk. You can use this image.

■ **Tip**   If you're enjoying this detour on the way to analytics bliss, Chapter 2 of my *Introduction to Social Media Marketing* book covers a few basic tools for working with content, including images, and source of images.

When I tried a slightly larger image, it still didn't work:

In the end, I just grabbed that "Earth" image, which was 300x300, and that worked.

After you have an image in place, type in a headline and text for the ad. It's an art and a science, but just put in something short and direct with a call to action. The little numbers on the right tell you how many characters you have left to type in. There are limits.

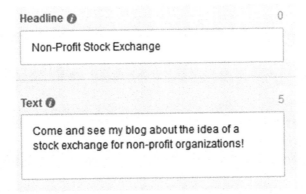

Next, you pick an audience. You could send it to completely random people, but you'll end up with more clicks if you indicate some interests.

Try clicking in the Interests field and typing in an interest, such as "social media marketing", "web analytics", or whatever your blog is about. If you don't have a topic, make it about online marketing, read all the free books, and keep posting!

As you type, options will come up and you can click on them:

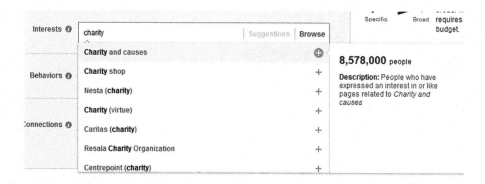

Narrow things down a bit and then review the budget at the bottom:

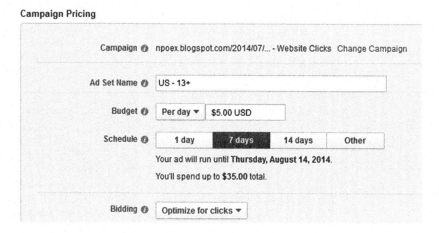

If you want to limit your investment, you can click on Per Day and switch to a lifetime budget.

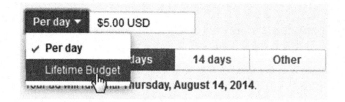

Then click on Place Order:

If you get a message like this, you might need to upload a new image:

Finish Your Ad

⚠ You are using an image that does not meet size requirements. Please adjust the size of your image or select a different one.

(You might need to click on the "X" by any image you uploaded that isn't large enough. You can use multiple images to see which generates the best clicks.)

You should get a confirmation, and you can click Continue:

Your Order Has Been Placed

Your order has been placed
You'll receive a notification once your ads are reviewed.

Learn about conversion tracking

Conversion tracking can help you better measure the performance of your Facebook ads by reporting the actions people take on your website after viewing your ads. Start Now.

If you want to come back later and see how things are doing in Facebook, try this link: https://www.facebook.com/ads/manage.

	Status ?	Campaign Name ?	Delivery ?	Results ?	Cost ?	Reach ?	Spent Today ?	Total Spent ?	Start Date ?	End Date ?
	⬤○	npoex.blogspot.com/2014 /07/... - Website Clicks	○ In Review	0 Website Clicks	--	0	$0.00	$0.00 of $34.93 ⓘ	08/07/2014 8:14am	08/14/2014 7:55am

(If you want to know more about Facebook ads, check out my *Introduction to Social Media Marketing* book).

The overall point is that a Facebook ad may generate more clicks and attention than your personal social posts. The more clicks, the more data you have for this learning experience. All the better.

And if you plan on running a month-long or quarter-long experience (Q1, Q2, etc.—three months), or even a semester-long experience, you can always set a monthly budget and an end date, and just let the thing run. If you do something like that, be sure to blog regularly, such as once a week. Follow the tips to add "Follow by Email". See how things go. In some cases, web analytics is a job unto itself, but it's fair to say that in many cases it's a skill anyone can learn, including people who are responsible for getting the traffic in the first place.

Learning More

Here are a few more resources for getting more traffic/contacts:

- Web Traffic 101: How to Get More Customers – Fast: http://blog.leadpages.net/web-traffic-101-how-to-get-more-customers-fast/

- 10 Ways to Get Traffic for Free!: http://www.webconfs.com/how-to-get-traffic-article-30.php

- How Do I Import Friends from Other Accounts?: https://www.facebook.com/help/189976267783273

- "How to Import Contacts into Twitter: http://www.dummies.com/how-to/content/how-to-import-contact-lists-to-twitter.html

- How to import contacts into LinkedIn: https://www.linkedin.com/help/linkedin/answer/4214?query=import%20contacts%20into%20linkedin

Conclusion

Congratulations on exploring how to generate analytics! It's important to set these things in motion, partly because having real, live data is more fun than just looking at a screenshot. Next up, we'll look at how to review performance.

Reviewing Performance of Campaigns

This chapter takes a look at some of the ways you can track the performance of a web site or campaign, as well as some related terms, and covers the idea of ongoing reporting. If you're following along in text only, you can pick up some information just from reading the material. However, I think it is more compelling if you develop a blog or site as discussed in Chapters 1-3, so that you have your own personal data to look at. It's more exciting, and even fun!

Reviewing Campaign Analytics

Even though our focus is on Google Analytics, I think it's worth seeing how tools relate. In the previous chapter, we looked at how to generate some traffic for viewing in analytics, and this would be representative of if you were doing an ad campaign and looking at its impact. In many cases, you could be involved in the execution of an ad campaign, as well as looking at the data.

It's also helpful to see how an ad platform views things, versus how things look on your actual web site or blog.

© Todd Kelsey 2017
T. Kelsey, *Introduction to Google Analytics*, DOI 10.1007/978-1-4842-2829-6_4

In other words, the clicks from an ad don't tell the whole story. Without looking at Google Analytics, you can get some sense of the amount of traffic you'll get, but you won't really know what to do with it. It's like sending someone to a house but not knowing what they do once they get there.

If you're following along and have tried a Facebook ad campaign, you can go to Facebook to see how things are going: https://www.facebook.com/ads/manage

At this end, you're seeing the "outgoing" information—people on Facebook who see your ad. Hopefully some of them click on it and visit your blog or web site:

	Status ?	Campaign Name ?	Delivery ?	Results ?	Cost ?	Reach ?	Spent Today ?	Total Spent ?
		npoex.blogspot.com/2014 /07/... - Website Clicks	Active	32 Website Clicks	$0.58 Per Website Click	9,395	$0.53 of $6.00	$18.81 of $34.92

If you tried the promoted post idea, you can go back to it and click on the Sponsored link. At present, Facebook doesn't give good data on personal sponsored posts. (Is this because it's in their best interest not to give numbers?) All they give you is a relative picture:

Okay, so how many people actually clicked on the link in the Sponsored post? This is one limitation of the "personal promotion" platform—limited data. When you run Facebook ads (or sponsored posts from a Facebook page, etc.), you get a bit more data.

Ultimately, you can look at Google Analytics to see how many clicks arrived from various sources.

Reviewing Performance in Google Analytics

So let's go back into Google Analytics (assuming everything is set up, you're tracking visitors, you shared on social media and/or ran an ad, etc.):

Go to `http://www.google.com/analytics/`. You should see something like this, where you click on the link next to the little globe:

Google Analytics has a lot of options—a lot of ways to "slice and dice"—but my recommendation is to approach it pragmatically. You don't have to know and use every option from day one. Instead, you can start by answering a specific question and branch out from there.

The fun part at this stage is seeing that there are in fact visitors to your site.

The Overview tab will open, and by default it will look back at recent data, based on Sessions:

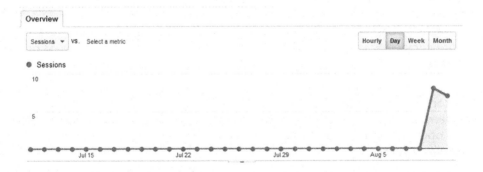

A *session* is a technical term for a visit, and the thing to keep in mind is that people often view several pages during a visit.

On the same page, there will be several additional items:

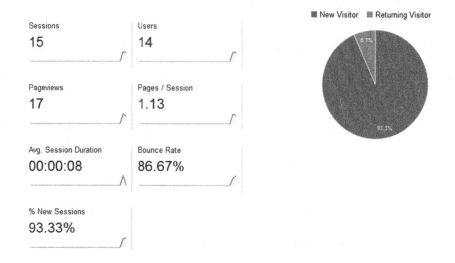

While there might be 15 sessions, there were 17 page views. This means that some of the visitors viewed more than one page during their visit.

Another interesting metric is the amount of time on average users spent during their visit. In this case, the average was 8 seconds.

One of the ways that people use information from Google Analytics is to view how performance currently is, and then set goals for the future. For example, a goal might be to look at the site or blog and see if you can add more compelling content, so that over time, the average duration of a session increases.

Bounce rate is a similar metric. The bounce rate represents the number of visitors who enter the site and then leave (or "bounce"), rather than continuing to view other pages on the same site. Another goal of reviewing performance and optimization might be to take note of the bounce rate, and see what you can do to decrease it.

What might be some reasons people "bounce"? A good Google search drill might be to search for "improving bounce rate" or "why do people bounce from a site"?

Location, Location, Location!

Another thing I think is particularly fun to do is to determine where people are visiting from. In Google Analytics, on the left, click on Geo:

Then click on Location:

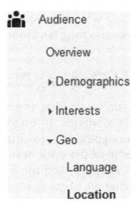

There is a map overlay, and it will give you an instant visual sense of where your visitors are coming from:

Then there is also a breakdown of which sessions were from which areas:

Country / Territory ⃝	Acquisition			Behavior			Conversions
	Sessions ⃝ ↓	% New Sessions ⃝	New Users ⃝	Bounce Rate ⃝	Pages / Session ⃝	Avg. Session Duration ⃝	Goal Conversion Rate ⃝
	15 % of Total: 100.00% (15)	**93.33%** Site Avg: 93.33% (0.00%)	**14** % of Total: 100.00% (14)	**86.67%** Site Avg: 86.67% (0.00%)	**1.13** Site Avg: 1.13 (0.00%)	**00:00:08** Site Avg: 00:00:08 (0.00%)	**0.00%** Site Avg: 0.00% (0.00%)
1. ▦ United States	**14** (93.33%)	92.86%	13 (92.86%)	85.71%	1.14	00:00:09	0.00%
2. ▦ Cyprus	**1** (6.67%)	100.00%	1 (7.14%)	100.00%	1.00	00:00:00	0.00%

For example, I'm not entirely sure why there was a visitor from Cyprus, but they didn't appear to stay on the site too long (and average session duration of 0).

Using Campaigns and Ongoing Blog Posts

With analytics, it's probably fair to say that the best way to learn the value, and learn how to optimize, is alongside an ongoing campaign of some kind, and/or a web site that has an ongoing set of visitors. If you're just starting out, exploring and playing with some of the options in Google Analytics can teach you a few tools that you can keep in mind for when you're in that situation. I personally believe that making an ongoing blog is a good general learning experience, as well as a source of traffic.

Dare I say that you can think of analytics as a party? Invite people from all over the world, give them something enjoyable, meaningful, or relevant to experience, and then see what they do. That's analytics!

Looking at Trends

One concept in analytics that's helpful to explore, file away, and burn into your synapses is the idea of *trending*. It basically means looking at how things perform over time. You look at the trends. An easy way to get a sense of trending is to go directly into Blogger and click on the Stats link. You'll see a simplified set of analytics.

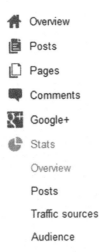

So the 2069.us blog has been around for a while. It is a personal blog that hasn't been promoted extensively, but has seen some traffic.

The Stats function in Blogger opens up to the Week tab:

If you click in the All Time tab (upper right), it shows a longer view, and the longer your blog has been around, the more you can see trends. In theory, with a consistent application of effort (adding quality content, making some effort to promote, getting listed on some other sites), you will see a boost in the amount of traffic.

On the left, since 2006 that is, we see that there has been a fairly gradual growth in the amount of traffic. This is a good trend—at least it's not going down.

Dealing with Spam Traffic and Bots

One issue to be aware of when monitoring traffic to web sites and blogs is that sometimes there are automated *bots* that appear like regular users. In theory, this traffic is filtered out, but part of the point is to think critically and dig beneath the surface. In Blogger, you can look at traffic sources, which is also an interesting way to learn more about your web site. You can see where your visitors are coming from.

For example, it appears that a large proportion of traffic is coming from semalt.com. What's that?

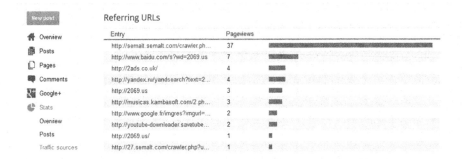

It looks like it is some kind of web promotion service. It's not a news site or blog. It looks like some kind of automated crawler that would get noticed by an analyst.

The longer a web site or blog is around, the more some of these traffic sources will result in true referrals—where people linked to your web site or blog because they like it. The idea is to keep an eye on traffic sources. In some cases, you might be able to establish a relationship with the traffic source and let them know when you have a new blog post or product.

As it is, because 2069.us hasn't been updated in a while, it looks like it's getting mostly automated traffic. Another traffic source is baidu, a Chinese search engine. I have no idea how the blog ended up on the site, or why people would visit it, but maybe it was just interesting enough to generate a few clicks.

Looking at Traffic in Google Analytics

You can also look at traffic in Google Analytics (in a deeper way). Try clicking on Acquisition ➤ All Traffic:

At this stage, I just recommend exploring and trying things out. There will be plenty of time to dive into documentation and details later. For now, try to treat Google Analytics like an amusement park, where you try to find things that are amusing. Who said data can't be fun?

Tracking Mobile Hits

If you were starting to doze off in the previous sections, maybe the idea of tracking mobile hits will get your attention. It's a hot, growing wildly successful area in consumer electronics, apps, and web site development. Analytics gives you a way to look at mobile hits. For example, in Google Analytics, go to Audience ➤ Mobile:

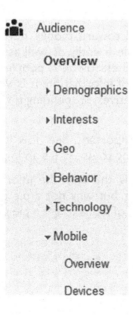

It can be interesting to see the percentage of your traffic coming from different devices:

Device Category ?	Acquisition		
	Sessions ? ↓	% New Sessions ?	New Users ?
	15 % of Total: 100.00% (15)	**93.33%** Site Avg: 93.33% (0.00%)	**14** % of Total: 100.00% (14)
☐ 1. desktop	**11** (73.33%)	90.91%	10 (71.43%)
☐ 2. mobile	**2** (13.33%)	100.00%	2 (14.29%)
☐ 3. tablet	**2** (13.33%)	100.00%	2 (14.29%)

This represents another way that analytics can be used to deliver value to a business. Do they know the percentage of traffic coming from mobile? What is the trend? Is mobile traffic increasing? This might mean that the business or organization needs to pay more attention to being mobile friendly.

If you start making your site more mobile friendly or use mobile advertising, analytics can help you measure the impact. The ROI. Remember that ROI is the Holy Grail!

You could look at the trend toward mobile device usage. You can look at it before your site was mobile friendly, as well as after. Is there a difference? Hopefully there will be. Business owners, people in organizations, pay very close attention to this kind of information. It is very, very valuable. You can see clearly the impact of initiatives, of spending time and money on campaigns and improvements.

This is why analytics is so important. See how it works? Just something to keep in mind as you're learning to use it. It's an important piece of the pie.

Another issue to consider is the way the information comes across. For example, a chart is one thing, but why not a pie graph? In Google Analytics, you can click in the little row of icons to get a pie graph in this view of mobile:

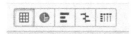

It opens a color-coded view that is easier to consume.

At a single glance, you see that overall mobile traffic is a quarter of the total traffic (mobile devices and tablets, which fall into the mobile category as well).

Device Category	Sessions ▾ ↓	Sessions	Contribution to total: Sessions ▾
	15 % of Total: 100.00% (15)	15 % of Total: 100.00% (15)	
1. ■ desktop	11	73.33%	
2. ■ mobile	2	13.33%	
3. ■ tablet	2	13.33%	

As you begin to look at these kinds of things, you can start to ask interesting questions, in a kind of "mix and match" mentality.

For example, what is the bounce rate for mobile traffic to your web site? A higher bounce rate on mobile devices might indicate that the site isn't mobile friendly. But how do you check this?

I think it's important to learn how to learn, so I recommend starting by asking these types of questions. Then see if you can figure out how to look under the hood. You can also cheat by Googling. For example, try Googling "mobile bounce rate in Google Analytics".

Using Dashboards/Reports

When you're in the position of looking at web analytics for your own site or for someone else's, you can manually go into the tool and find items. However, this can be time consuming, and Google Analytics includes dashboards and reports you can create to make your life easier.

In Google Analytics, click on Reporting at the top:

Then choose Dashboards ➤ New Dashboard:

Then select Starter Dashboard and change Untitled Dashboard to whatever name you like:

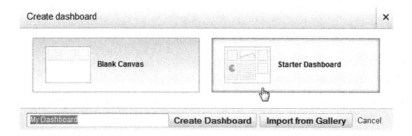

Finally, click Create Dashboard.

Next, you will probably want to set a date range. There's nothing wrong with the default, but setting date ranges is part of the way you get the information that's most relevant.

To change the date range, click on the little triangle to the right of the dates that show:

Consider the preset options, such as Last Week, which might be a good date range for a regular, weekly report:

After you select a date range, click Apply:

Exporting and Scheduling Reports

Next you may want to manually export a report, in order to share it with colleagues or clients. To do this, you click Export and select PDF:

I also recommend setting up a scheduled report via e-mail. To do this, click Email:

Then enter an e-mail address to send the report to:

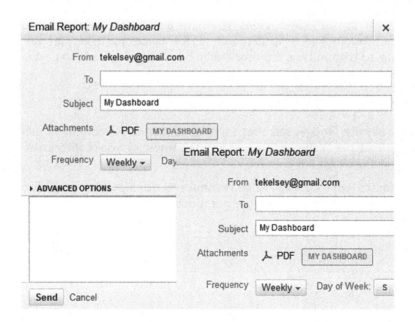

Then click in the main area and add a message, such as, "Here is the weekly report". Then click Send:

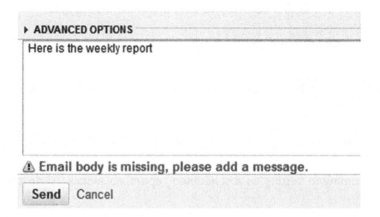

You can always come back to Dashboards to adjust your options:

If you went through these steps, you have fully entered the world of web analytics. You looked at traffic and performance and then set up some reporting so that you can monitor traffic on an ongoing basis. Woo-hoo!

Learning More

Aside from the Google searches mentioned earlier, I recommend exploring Google Analytics. When you come across something you're interested in, do a Google search on the concept.

To synthesize your learning, I also recommend making a blog post, discussing what you're learning, including the questions you have, and asking people for comment.

You might also be interested in a little light viewing/reading:

- Basic Concepts in Web Analytics: `http://ramonlapenta.com/blog/basic-concepts-of-web-analytics`

- A Web Analytics Slide Presentation: `http://www.slideshare.net/mattPR0v1/web-analytics-concepts-theories`

- Bounce Rate: `https://support.google.com/analytics/answer/1009409?hl=en`

Conclusion

Congratulations! You just made it through a significant set of milestones. You broke through the data barrier, created some traffic, learned how to view data using various tools, experimented with multiple social media channels, and looked at behavioral data. Just getting your feet wet in these areas is impressive. Good job!

If there's any way for you to have the right mindset that will help you get through the material and master the tool, I do believe that *fun* can pave the way. Hopefully, you can find something interesting when creating your own site, trying to promote it a little, and looking at how consistent effort over time can have an impact. I argue that web analytics is a critical part of marketing partly because of inspiration. That is, when you see the impact, the results of your effort, it inspires you to pursue things consistently over time.

This translates into fiscal ROI and an impact in a business setting. When you trace impact, you can show how important it is to make adjustments and show the importance of the online marketing efforts. It really helps, and it definitely gets attention. Instead of guessing and hoping, you get a clear sense of what's going on.

Fun with E-Commerce Analytics Part I: Shopify

Tracking ROI is a 50 billion dollar skill, because it's at the core of Google's success, and you could even argue that it's a trillion dollar skill, because of how much revenue Google helps businesses make. This chapter looks at how to set up a Shopify account, which is one of the easiest ways to set up a "real" e-commerce system to learn about tracking ROI with analytics. I think it's important to see how you can set things up so that when you create an ad, you can track exactly how much money you are making. This is a fundamental concept and opportunity in analytics, where Google Analytics and AdWords can act together. My goal has been to find the easiest, least expensive way to set up the moving pieces, and Shopify is part of the puzzle. Originally, I was going include the entire lifecycle in one chapter, but I decided to split the chapter into two—this first chapter lays the foundation in Shopify, and the next chapter covers connecting things in AdWords.

© Todd Kelsey 2017

T. Kelsey, *Introduction to Google Analytics*, DOI 10.1007/978-1-4842-2829-6_5

In an ideal world, this would be free, but from what I can tell, there isn't an e-commerce platform out there that is free and allows you to do conversion tracking. At the time of writing, there are some platforms such as Gumroad (covered in Chapter 7), which is free and provides some basic tracking, but if you point an ad at a Gumroad site, you are still guessing about the effectiveness of particular ads.

The great strength of being able to do "real" conversion tracking is that in Google AdWords, you can tell exactly how effective a particular ad has been, in terms of revenue. When you or your employer or client is spending money on ads, it's important to know what is most effective, so you can spend money on ads with confidence, instead of through guesswork. That's the exciting thing about setting up the full deal.

Setting up Shopify, at $29.95/month, is about the best and easiest way to do things at present (e-mail me if you know of an alternative).

Understanding Conversion Tracking

At a high level, conversion tracking allows Google to make 50 billion dollars a year. The magic behind this machine is that unlike most other forms of advertising, with AdWords you can get *analytics* that tell you exactly how much money you are making, based on what you are spending. If you spent money on a billboard, you can only guess at how many people are influenced to purchase—and the same thing goes for television, radio, and most other forms of advertising.

If you're spending $1000/month on Google ads and you know you're making 10,000 as a result, you can confidently consider increasing the budget, or just keeping it in place. Without the conversion tracking, you have no idea. Of course, selling something online is not as simple as just placing an ad on Google. People have to want what you are selling, and there's also *competition* for Google ads. But when you follow the best practices for online marketing, and when you have conversion tracking in place, it gives you analytics to track ROI—the bottom line.

In terms of Google Analytics and Google AdWords, they are both tools. They work closely together, but it's the conversion tracking that makes the difference.

Why Shopify? Why "Live"?

My recommendation is to follow along in this chapter as a learning experience about how e-commerce comes together, to set up a "live" Shopify site. Start thinking about a product that you could sell, even a physical product if you have one, or think of someone who might like some help exploring this area. Another area that is slightly easier to develop and manage, as part of a learning experience, is a *digital* product. Take anything that could be written/developed in Microsoft Office or Open Office, save it in PDF format, and make it a digital download.

You are also welcome to use the *Introduction to Social Media* book PDF, or one of my other books, as a live file to place on your "test" Shopify store. Ultimately, the competition for keywords about social media marketing in AdWords means the money you have to spend in order to actually sell the PDF might be more than you want, but it's still worth the test. The only thing I'd ask is that you contact me first, partly because if multiple people are using it as a sample, it makes sense for me to coordinate "areas" that you could target your ads to.

Don't worry if you don't have experience in e-commerce, AdWords, etc. I try to make it as easy as possible to start out. Even if you want to keep spending down to a bare, bare minimum, you can get one or two friends to search on Google for the keywords you choose, click on the Google ad, follow it through to your site, and purchase the item. That will allow you to see the full lifecycle of conversion tracking.

It is an exciting thing, to be able to trace your effects for real. It's also an important skill, and I do think that it's worth pursuing, so that whether you are just learning for yourself, expanding your skillset at work, or building your skills to find work, you can say "I have experience setting up an e-commerce site and generating analytics with conversion tracking in AdWords". This will definitely get people's attention.

There might be alternatives to Shopify, but Google itself recommended it as an option for e-commerce when they discontinued Google Checkout, and Shopify has pretty good support. Using a combination of these chapters, the extra links for learning material, and the 800 numbers for support, I think it is realistic and doable to try this out. There's also the Shopify LinkedIn group to pose questions to as well (ideally after you've tried asking either Google or Shopify support).

It is doable, and you will find it exciting, so I recommend trying it. First read both of these chapters and then try creating a Shopify site. Then set up the Google AdWords after that.

Think of it as a ride at an amusement park—it will be a new experience—but it can be fun!

Get Started with Shopify

To get started, go to shopify.com and click Get Started:

■ **Note** Until/unless things change, you can definitely do a free trial to try Shopify without spending money. In order to be able to do the conversion tracking, though, you have to pick a plan and spend $29.95/mo or so until you cancel. Given the impact it could have on your business, career, and employer, I think it's worth it.

Next, enter your e-mail address, a password, and a name for your store. Click Create Your Store Now:

The store name could be anything you like it to and should ideally be related to the product you are selling. To make it easy, you could just use your initials, and say something like JDmarket. You'll end up with an address like jdmarket. shopify.com, and if you like this e-commerce thing, you can always upgrade to a full web site name like www.jdmarket.com.

When you've got a name, choose the Online Store checkbox and click Next:

Where would you like to sell?

Next

Then enter your name and address information:

Add an address to set up currencies and tax rates

· ● ·

FIRST NAME

LAST NAME

STREET ADDRESS

CITY

ZIP/POSTAL CODE

COUNTRY

United States

STATE

Click Next when you're done:

At this point, I'm going to follow the scenario of digital products. If you're setting up your site to sell physical products, there's more to configure, but you can always guess and then call Shopify support if you need to.

You make some basic choices and click Take Me to My Store:

Tell us a little about yourself

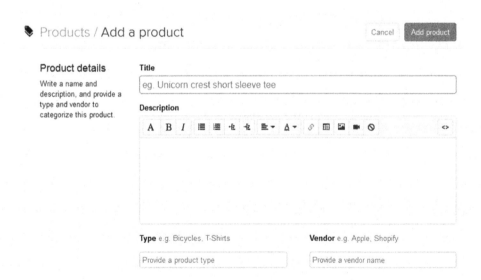

Next, you add a product.

If you are following along with a digital product, you can enter a title.

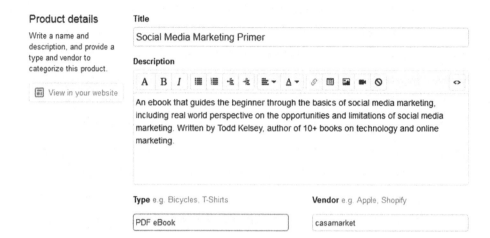

Then click Add Product:

Next, you choose images (such as a book cover for a digital product). If you are uploading your own and don't have a cover image, try hiring someone on elance.com or odesk.com or read the "Content" chapter in my *Introduction to Social Media* book for some tools on working with images. Try for a 250x450 image using SnagIt, Photoshop, or Gimp. You can add text and use clipart.

Images

Upload and edit images of this product. You can also add images from the web. Drag to reorder images.

→ Use the **Choose images** button to add images to this product.

Choose images

After you upload an image, it should look something like this:

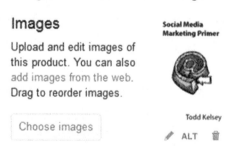

Then click Save:

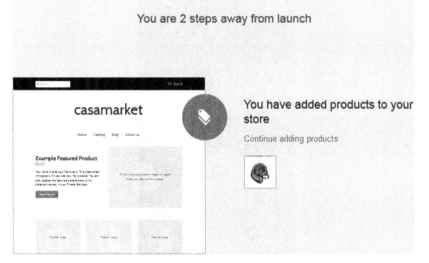

The Shopify wizard will walk you through things, and you can click on the 2 steps link:

In theory, you should also see a link like this, where you can click View Your Store to see a preview of what it looks like:

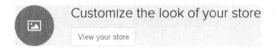

Customize Navigation

The next thing you'll want to do is to customize navigation. If you know that only you and your friends will be looking at the store, not all the steps I go through are necessary, but it's still a good experience to do it. The more likely you are to try and actually get people to visit the store and buy something, the more you'll want to tweak your site.

To customize the navigation, go to the menu on the left side of the screen click Navigation:

In this section, I recommend taking out a few things, such as the link to the blog, so you don't have to add content to it or maintain it. If you want to blog about your e-commerce store, go for it! Just make sure you regularly add posts. You don't want potential customers to come to the site and find old blog posts, or no blog posts, which will make your site seem less professional. You can also add a blog later.

Next, the Edit Link List link:

Then click the trash icon next to the blog:

Finally, click Save:

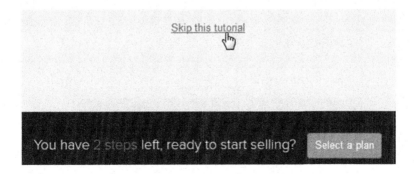

If you encounter a link like this, you might just want to skip it for now:

Adding the Product to the Front Page

Even after you add a product, you'll want to add it to the front page. To do this, go to the left navigation and click Collections:

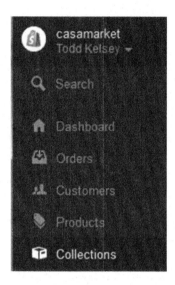

Look for the Front Page collection. Then click on the Add Products menu:

Products

Manage the products that belong to this collection.

Select the product you added to the system:

It should show up in the list:

Products

Manage the products that
belong to this collection.

Social Media Marketing Primer ✕

Then you can click Save:

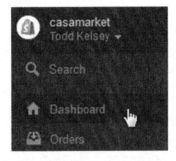

Tweaking and Payments

Shopify makes it about as easy as possible to set up a fully functional
e-commerce site and accept payments. It used to be, and often still is, the case
that you needed to be a web developer, hire one, or learn how to be one, and
then work with a hosting company to put a lot of pieces together, including
adding payment processing to your web site, etc. Suffice to say that jumping
through the hoops on Shopify is a lot easier than other setup situations.

Next, go to your Dashboard:

You will probably see a wizard like this:

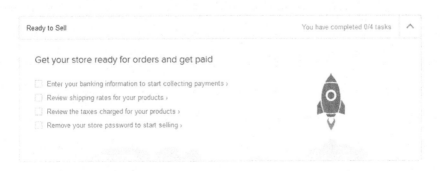

If you're following the "digital products" example, all you need to do is enter your banking information. Even if it's just you and your friends, colleagues, or classmates testing things out, you should enter the information to get the full effect.

First, click Complete Shopify Payments Account Setup:

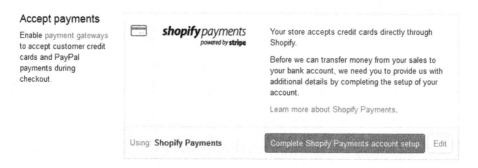

You'll probably want to choose Individual/Sole Proprietorship:

Your business type

Individual / Sole Proprietorship ▾

■ **Note** By continuing to read, you are accepting the fact that I am not giving you tax or business advice, just options to consider.

Next, uncheck We Sell and Ship Physical Goods:

Type in eBooks instead:

What kind of products or services will you be selling?

eBooks

Then enter your business address:

Finally, enter your bank information:

Click Complete Account Setup:

Complete account setup

Then click Save:

🛒 Payments   Save

Your store accepts payments with: PayPal Express Checkout, Shopify Payments.

Then click Back to Admin:

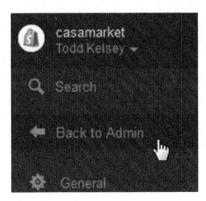

Next, click the check boxes on the left, to set up shipping rates and tax rates:

Get your store ready for orders and get paid

☑ Your bank account is now set up to receive payments ›

☑ Your shipping rates are ready for your first order ›

☑ Your tax rates have been set ›

☐ <u>Remove your store password to start selling ›</u>

Then click Remove Your Store Password, and it will lead you to pick a plan, which is covered next.

Picking a Plan

This is where you have to pay. The trial allows you and others to visit the site, and it's live, but they need a password, and random visitors from a Google ad can't get in. In *theory*, if you tell friends to find the Google ad and click on it and you give them the password, technically you might be able to try Shopify for free. I haven't tested that idea, but it's possible in theory.

My recommendation is to make it live. Go through the exercise of making it as professional looking as possible, spend the money if you can, then find a person with a real product, physical or digital, to sell. Chances are there are a fair number of people in your community who might be trying to find a solution like that, and if someone comes along to help them, they might even pay for everything, as well as pay you to help them.

Click Pick a Plan:

Storefront
password

You can protect your store
with a password and show
a customized message.

 Your store is password protected with the password: **glaomp**. To remove your
storefront's password pick a plan.

I suggest the cheapest plan. Click Choose this Plan:

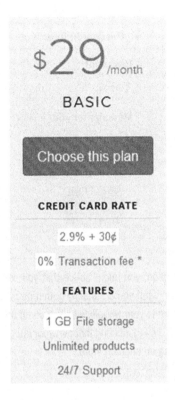

Start with the Once a Month plan instead of the yearly plan:

Billing cycle	● Bill me once a month for $29.00

Then click Confirm changes:

Confirm changes

Next, click Remove Your Storefront Password:

🛒 **You're almost ready to start selling!** Once you remove your storefront password, your store will be live and people will be able to buy your products

Uncheck Password Protect Your Storefront:

Then click Save:

Note If you are paying attention, you might notice that you can connect Google Analytics to your storefront. When you do that, what you are mainly gathering is "conventional" analytics, such as behavior and number of visitors to your Shopify site. You could try that, but the focus for now is on setting up conversion tracking. I suggest filing that away for later, if you plan on wanting to learn about e-commerce analytics using an ongoing store.

Click Back to Admin:

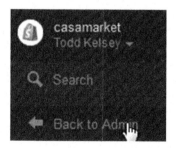

Setting a Price

For any product you sell, you'll need to set the price. Click on the Products link:

Then select a product to price:

		Social Media Marketing Primer	N/A	PDF eBook	casamarket

In order to have a price, you have to make a "variant". E-commerce items like t-shirts will often have different sizes, colors, etc., so for a digital download, you could just have a name for a particular edition, like "First Edition" or "Version 1.0".

To proceed, click the Edit link:

Inventory & variants	Title	SKU	Price
Configure the options for selling this product. You can also edit options.	Default Title	—	$0.00 Edit

Add a variant

Type in a variant name and price:

Title

Version 1.0

Price

1.00

■ **Note**   By continuing to read, you agree I'm not giving you business or tax advice.

If you're doing this as a limited time experience, you may not need to charge taxes, and unless you are shipping physical goods, you don't need to require a physical address:

☑ **Charge taxes on this product**

☑ **Require a shipping address** (not needed for services or digital goods)

So uncheck both of those boxes and click Save and Close:

☐ **Charge taxes on this product**

☐ **Require a shipping address** (not needed for services or digital goods)

Inventory policy

| Don't track inventory | ▾ |

Cancel   Save   Save and close

Tip   If you plan to sell things on an ongoing basis, I recommend finding an accountant. The old principle is, focus on what you do the best (such as marketing) and work with others to do the rest (such as accounting). You can do preliminary research on Google, such as "do selling digital downloads require charging taxes?". You can also call Shopify support for questions. Most likely they will help you configure your site, but will probably say that ultimate tax strategy should be determined in conjunction with an accountant. At a high level, the landscape seems to shift. In some cases, if you're shipping to the same area as your business, taxes may apply, and because of changing legislation, there may be taxes anyway. For a long time businesses like Amazon didn't have to charge taxes for online purchases, whereas buying in a local retailer would cause local sales tax. But local and state governments are catching up with this and starting to push for tax revenue. Something to research if you're interested in e-commerce. Ultimately, I recommend working with an accountant. Better to do it right than get burned later.

Next, you might like to scroll up on the screen and click View in Your Website:

Product details

Write a name and description, and provide a type and vendor to categorize this product.

 View in your website

Then take a look. Congratulations! You have fought half the battle in setting up an e-commerce site. In the next chapter, you'll get everything configured and find some visitors.

Home   Catalog   About Us

Home > Products > Social Media Marketing Primer

Social Media Marketing Primer

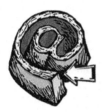

Todd Kelsey

Social Media Marketing Primer

$1.00

Version 1.0 ▾

Add to cart

An ebook that guides the beginner through the basics of social media marketing, including real world perspective on the opportunities and limitations of social media marketing. Written by Todd Kelsey, author of 10+ books on technology and online marketing.

Share this item: 8+1 Tweet Pin it
Like ⟨ 0 ⟩

The link for your site will be something like this: `https://casamarket.myshopify.com`.

Making More Tweaks

These tweaks are not strictly necessary, but recommended. Even if you took the blog link off, there might still be a News section on your front page:

News	Quick Links	Follow Us
First Post	Search	
This is your store's blog. You can use it to talk about new product launches, experiences, tips or other news you want your customers to read about. You can check...	About Us	

To get rid of this, you can click Themes:

Then click Customize Theme:

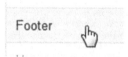

And Footer:

Then click the X next to blog:

Note This page also happens to be where you can configure "where" your social media icons point. I wouldn't worry about it, but you can always come back later and add this. In the meantime, if anyone clicks on those icons, they will be taken to Shopify.

Social Links

Twitter   http://twitter.com/shopify

Facebook   http://facebook.com/shopify

Finally, click Save:

Updating Content

You'll want to spruce things up a bit. Click on Pages:

And then click Frontpage:

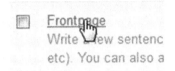

Add some descriptive text:

For example, using the sample text/files mentioned earlier, you could add the following and click Save:

(If you want to "go pro," hire a copywriter, locally or through elance or odesk.)

Next, click Pages:

Click on the About Us link:

Using the sample text or your own text, enter the About Us text. You can use Shopify's suggestions as a guide:

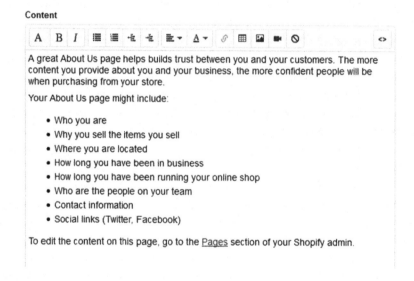

Feel free to use the sample text as well:

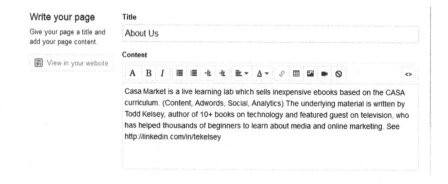

Then click Save:

Adding the Digital Product

Next, you'll need to upload a digital product, using the Shopify app.

Click on Apps:

Then click Visit the App Store:

You don't have any apps installed

Visit the App Store

Type in digital downloads and select from the drop-down list:

Next, click Install Digital Downloads:

Now you need to select a product that you set up before. Click the Product link:

You don't have any products with file attachments.

Select a product in your Shopify admin and click the *"Add Digital Attachment"* button to begin to add files.

Then select your product you want to upload:

	Product ▲
	Social Media Marketing Primer

Click on the "…" icon and choose Add Digital Attachment:

Next, click Upload File:

The file should begin to upload:

Version 1.0

$1.00

◎ Uploading: Social Med...Primer.pdf at 701 kb/s   **UPLOAD FILE**

Congrats! You set up a digital download.

To see your store, click on the middle icon at the bottom of the left navigation:

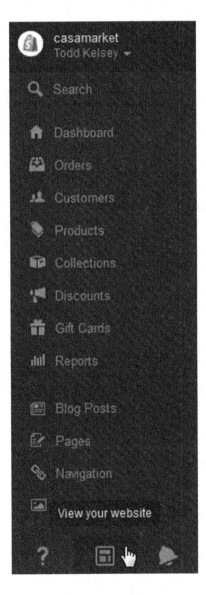

Buy Your Product

At this point, I recommend buying your own product, just to test it out and get that sense of satisfaction that comes from setting things up. Click Add to Cart:

Social Media Marketing Primer

$1.00

Version 1.0 ▼

Add to cart

An ebook that guides the beginner through the basics of social media marketing, including real world perspective on the opportunities and limitations of social media marketing. Written by Todd Kelsey, author of 10+ books on technology and online marketing.

Share this item: 8+1   Tweet   Pinit   Like {0}

Click at the cart icon at the top of the screen. If you want to make it clearer to visitors what to click on and where, you can add text to your product description, such as "Click on Add to Cart to buy, then click on the Cart icon at the top of the screen to check out."

Next, you can click Check Out:

Your cart

		Quantity	Price	
🖼️	Social Media Marketing Primer	1	$1.00	Remove
			$1.00	

Add special instructions for your order...

Update   Check out

Check out with **PayPal**
The safer, easier way to pay

You'll get the standard user experience. For digital downloads, it collects an e-mail:

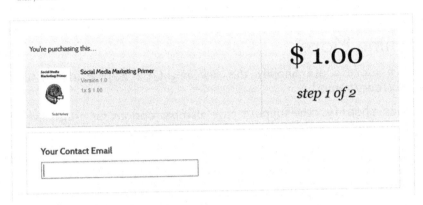

Note   If you plan to do this on an ongoing basis, you'll want to establish a privacy policy and put it on the site in relation to e-mails you capture. Ask Shopify about this.

Next, you'll get a confirmation, and buyers can download "now" or through the e-mail they receive.

Your Order ID is: **#1001**

An email receipt containing information about your order will soon follow. Please keep it for your records.

You will also receive an email with download links for your digital purchases.

Social Media Marketing Primer - Version 1.0

Social Media Marketing Primer.pdf (9.07 MB)

Download Now

Thank you for shopping at casamarket

Take a Deep Breath

Okay, take a deep breath.

Congratulations! You made it through setting up an e-commerce site. It's live and ready to advertise!

Learning More

To learn more about Shopify, this link might be helpful: http://docs.shopify.com/manual.

The docs.shopify.com/support page also has options for e-mail, live chat, and calling 24/7.

You can also get assistance from a Shopify expert, in hiring someone to help you set things in motion, via http://experts.shopify.com.

Conclusion

I think that going through the process of setting up an e-commerce site gets at the heart of analytics, bringing you closer than ever to being able to experience ROI first hand. Not many online marketing professionals can say they've done this, especially on their own. Even if you don't plan on selling anything yourself, I recommend considering this. If you end up with an employer or client who is trying to figure this out, you're one step ahead. You might end up doing analytics in a situation where all the underlying connections have been made, and I still recommend learning about this, so that you can have the "Aha!" moment. It's partly about inspiration. I do believe that ROI is one of the most important concepts, if not "the" most important concept, in online marketing. This is true whether you see analytics as your full-time job or just a skill in your toolbox.

Getting your head around ROI, really understanding it, will help you to be a stronger analyst and online marketer, period. If your emphasis is on ROI, you will be helping yourself, your employer, and your clients create more sustainable online marketing efforts.

Fun with E-Commerce Analytics Part II: AdWords

The purpose of this chapter and of Chapter 5 is to capture the full lifecycle of analytics and take a close look at ROI (return on investment). The goal is to shed light on an elusive question—when you are spending money on ads, how can analytics help you determine how much money you are making? The exciting thing is that AdWords provides a way to do this.

Shopify provides a foundation for e-commerce, and AdWords provides a way to advertise a site and track conversion. If you haven't read Chapter 5 yet, I recommend starting there. If you haven't read Chapters 1-4, I recommend starting at the beginning.

E-commerce and AdWords are billion dollar industries, but with the right approach, anyone can explore this world in a microcosm. As part of learning about analytics and specifically about Google Analytics, it's worth the effort.

© Todd Kelsey 2017
T. Kelsey, *Introduction to Google Analytics*, DOI 10.1007/978-1-4842-2829-6_6

How does AdWords relate to Google Analytics? When you're in an e-commerce situation, you can use Google Analytics to track behavior on your shopping site, just like you can track any other kind of site, in terms of visits, how much time people are spending on the site, etc. Platforms like Shopify can connect directly to Google Analytics for this kind of information. But what AdWords brings to the table is the ability to track *conversions*—measuring the journey customers take from clicking an ad to making a purchase. This is why AdWords is so important in generating information and why search engine marketing is a desired skill, and important for online marketers to understand.

If you want to learn more about SEO, check out my book titled *Introduction to Search Engine Marketing and AdWords* (Apress, 2017).

Okay, let's dive in!

Get Started with AdWords

To get started with AdWords, you can follow along in this chapter, if you want to dive right in. I have covered the bases as best I can, in a way that allows you to move along and try it directly.

If you want some background information first, there are also resources you can review, which are also included at the end of the chapter in the Learning More section.

- www.google.com/adwords for general information and links for learning more

- https://support.google.com/adwords is the learning center, including guides

If you come to see the value of having AdWords as a skill, I encourage you to explore Google's learning material and consider getting AdWords Qualified. It's free and it can help your career in a direct way.

To get going, the first step is to create an account. Woo-hoo!

Tip Free AdWords credit. If you've been following along and created a Shopify site in the last chapter, there are ways to get free AdWords credit, which is basically free money (or free ad budget at least). See:

http://ecommerce.shopify.com/c/ecommerce-marketing/t/aha-this-is-how-to-get-your-google-adwords-and-facebook-credit-121154

Creating an Account/Getting a Tracking Code

To start an account, go to `http://www.google.com/adwords`.

Feel free to wander around. The How it Works link may be of interest. When you're ready, click Start Now:

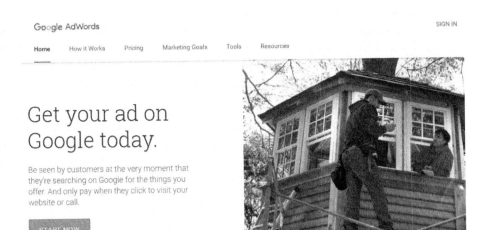

Note You may want to write down the toll-free support number (1-800-919-9922). It's in Google's best interest to help you succeed, and they have pretty good support.

When you have an account started, go to Tools ➤ Conversions:

Click Conversion:

This process is setting up a connection between AdWords and your e-commerce site. It results in a bit of code that you can bring into Shopify, which allows AdWords to track your site.

For example, the scenario you are setting up is that a person sees your ad, clicks on it, reaches your site, and ideally buys something. This "conversion code" you're creating allows Google to make a connection between the ad and your site. When the visitor reaches the Order Confirmation page or the Thank You page, the code is there. Google can report that this specific ad was clicked on and resulted in a purchase.

This basically allows you to determine that when you spend a certain amount of money on ads, it results in a certain amount of revenue on your e-commerce site. AdWords does not magically sell things for you. There's an art and science to selling, but the fundamental opportunity is significant for any business, and it's very solid, compared to traditional forms of advertising.

If you're planning on connecting AdWords and Shopify, you may want to review this link, which provides an overview and additional information around connecting the two:

`http://docs.shopify.com/manual/your-store/dashboard/google-adwords`

After clicking the Conversion button, enter a conversion name:

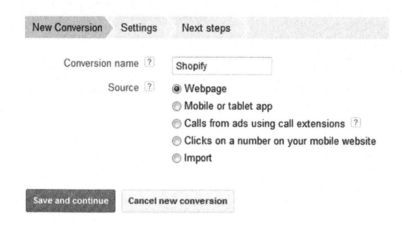

Click Save and Continue.

Then click "The value of this conversion action may vary". This allows you to set different prices.

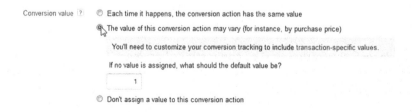

Next, set the Conversion Category to Purchase/Sale:

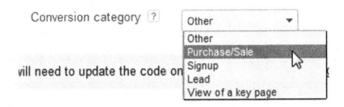

Make sure that the Markup Language is set to HTML:

Then click Save and Continue:

Save and continue

In the next section, click "I make changes to the code":

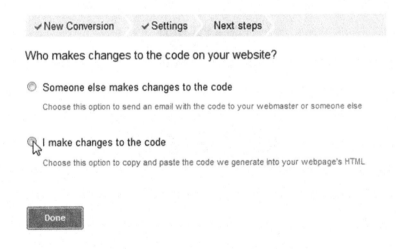

Finally, click the Done button.

Next, Google will give you the code, and you should copy it (Ctrl+C):

For example, paste this code into the webpage the user sees after signing up for your newsletter

```
<!-- Google Code for Shopify Conversion Page -->
<script type="text/javascript">
/* <![CDATA[ */
var google_conversion_id = 1044410043;
var google_conversion_language = "en";
var google_conversion_format = "2";
var google_conversion_color = "ffffff";
var google_conversion_label = "ntvnCPWE5goQu92B8gM";
var google_conversion_value = 1.00;
var google_remarketing_only = false;
/* ]]> */
</script>
<script type="text/javascript" src="//www.googleadservices.com/pagead/conversion.js">
</script>
<noscript>
<div style="display:inline;">
<img height="1" width="1" style="border-style:none;" alt="" src="//www.googleadservices.com/pagead/conversion
/1044410043/?value=1.00&amp;label=ntvnCPWE5goQu92B8gM&amp;guid=ON&amp;script=0"/>
</div>
</noscript>
```

Connect AdWords to Shopify

Back in Shopify, go into Settings:

Click Checkout:

Paste the code into the Additional Content and Scripts area:

Additional content & scripts

Any additional instructions or scripts you'd like to appear on the "Thank You" page of the checkout. This is an excellent place to paste things like ROI/conversion tracking codes and partner tracking systems.

Click the Save button:

To Modify or Not to Modify

In terms of actually changing the code, Google says you should do it:

You'll need to customize your conversion tracking to include transaction-specific values.

The Shopify documentation provides a way to do it (even showing relevant parts in red):

Manipulating the snippet to provide real data

In the code snippet you just pasted, **replace**:

```
var google_conversion_value = 1;
```

With this code:

```
if ({{ subtotal_price }}) { var google_conversion_value = {{ subtotal_price |
money_without_currency }}; }
```

If you are approaching this as a learning experience, you don't have to change the code, though. Just remember that when you try the ad out and get someone to click on it and buy your test product (even if it's a friend), you'll get a conversion value, but it will be $1.00. If you actually want it to represent the price of the products, you have to replace the code.

Don't be alarmed about the code; just think of it as a recipe.

For example, a recipe might say:

1) Get some bread

2) Put some peanut butter on a slice

3) Put the other slice on

Google is giving you a recipe for Shopify:

1) Here's some code

2) Shopify, the value we place on the items will be $1.00 unless we change the recipe

3) Have a nice day

It might look a little different (everyone has their own lingo, right?)

```
var google_conversion_label = "ntvnCPWE5goQu92B8gM";
var google_conversion_value = 1.00;
var google_remarketing_only = false;
```

You go to the Shopify article and select the relevant recipe replacement:

With this code:

```
if {{{ subtotal_price }}} { var google_conversion_value = {{ subtotal_price |
money_without_currency }}; }
```

And you replace the particular part of the "code recipe" with something new:

```
var google_conversion_label = "ntvnCPWE5goQu92B8gM";
```

```
if ({{ subtotal_price }}) { var google_conversion_value = {{
subtotal_price | money_without_currency }}; }
```

```
var google_remarketing_only = false;
```

If we bring it back into the kitchen, all you're really doing is saying that instead of peanut butter, you can choose what to put in the sandwich. So instead of this:

1) Get some bread

2) Put some peanut butter on a slice

3) Put the other slice on

You end up with this:

1) Get some bread

2) Put some [whatever you want] on a slice

3) Put the other slice on

All the code is doing is having one computer speak to another, or one program speak to another, etc. It's basically just a set of directions to do something.

Given the fact that you have the article, Google support, and Shopify support, I think you can do it. The advantage to changing the code is that then your site is real. That could be exciting, to know you've done it, and to be able to say you've done it.

But don't feel bad if it seems like too much—you can always come back later and try it for real.

Even if you don't modify the code, you'll still end up being able to test the connection between AdWords and Shopify, and all your wildest analytics dreams will come true.

What's Going On

To come back to Earth from the galactic analytics kitchen, consider the Order Confirmation page. The code that Google gives you, a "snippet" of code, which you can either use as is or tweak, is used by Shopify on the Thank You page.

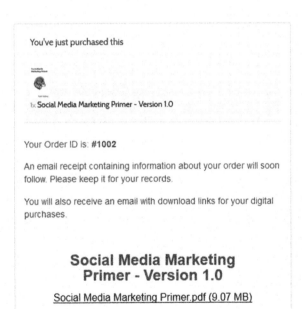

So when someone clicks on your ad, purchases something, and checks out, they end up on the Thank You page. When it displays, it tells Google there was a purchase.

▧ **Fun**   If you don't believe me, you can test the shopping cart process out (after you've added the conversion code). Then, in your browser, right-click on the page (Windows) or Ctrl+click (Mac), and choose View Source. Pretend you're looking at a recipe book and see if you can find something that looks like the conversion code that Google gave you. Then you'll win a prize! Okay, the prize is the knowledge that you've discovered the code you planted there.

Create an AdWords Campaign

Now that you've connected Shopify and AdWords, you can create an ad so that you can try it in real life.

Alternatively, you can get something a friend wrote and sell it as a digital product, or even find a physical product. But to keep the momentum going, all I'm saying is there are options.

In short, the world's your oyster!

What you'll see very soon is that AdWords ads cost money—surprise! But that's fair. In order to make money, you have to spend money. Analytics allow you to see if you're spending money effectively. Did your ad work? If people clicked on it, did they buy anything?

To be clear, Google charges you when people click on your ad. This provides you visitors. It's up to you to get people to buy something.

There's competition, so when you're paying Google for the clicks, it's like a kind of eBay auction. For example, say you own a company that sells basketballs and there's another company that does the same. You both want to get people to click on your ads. When people type "basketball" into Google, there's a bidding situation.

This is where it can be helpful to review some of the background info, as mentioned earlier.

On the bright side, you can sometimes get free ad credit. Check out:

`http://ecommerce.shopify.com/c/ecommerce-marketing/t/aha-this-is-how-to-get-your-google-adwords-and-facebook-credit-121154`

At this point, you can limit your "ad spend". You bid very low and make your product price very low, just to get some friends to find the ad, click, and buy the product, so you can test things out. Then you can always come back and try the finer points by experimenting with different prices and bidding, in an effort to get "real" customers. That can be exciting. Think of that—an exciting analytics learning experience.

Yes, it can take money, but it can be exciting. Especially if you're doing it for a client, a friend, or an employer. At that point, I definitely recommend getting AdWords Qualified, so that you can increase your chances of selling effectively.

At any rate, when you're ready to try AdWords, go into AdWords and click Campaign:

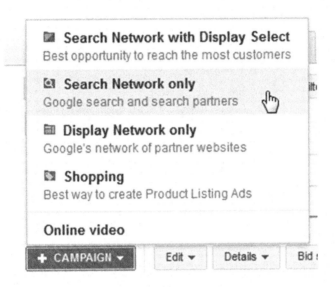

For the moment, click Search Network only.

Next, give the campaign a name:

Next you will be given a chance to set a budget. Ignore everything else for the moment.

I suggest $5.00 a day for learning. It doesn't mean you'll necessarily spend $5.00/day; this is only if people actually click on the ad. But you can limit a budget and change it later, and that's the important thing to remember.

Feel free to click on the little question mark icons wherever you see them to learn more.

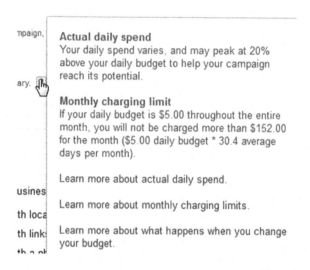

Then click Save and Continue:

Next, the wizard will ask you to enter your landing page:

Ad groups

Enter your landing page [?]   ☐ My landing page isn't ready

The webpage your ad will link to (ex: www.example.com/yoga)

Enter your landing page to get keyword suggestions

You can get this page URL by going into Shopify and clicking on the middle icon on the bottom-left side:

Technically, your site page can be your landing page. This is the page where people go when they click the ad.

As with other aspects of AdWords, there is an art and science to landing pages, which is part of the way you increase the chances of someone buying something (by selling the value proposition, etc.). Just for learning purposes, you can copy that link from your browser and bring it back into AdWords.

Next, you can give your ad group a name if you like. What AdWords is doing is giving you an opportunity (by scanning your "landing page") to get some ideas for keywords.

When creating an ad, ultimately you are targeting a particular keyword or set of keywords related to your product. Ask yourself, what would people type in Google if they were interested in my product?

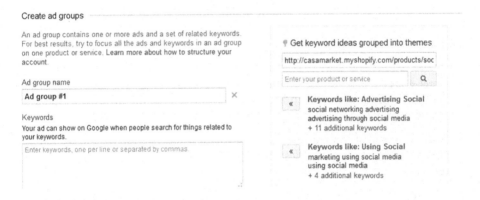

This screen gives you the ability to get some ideas. Click on the little arrow icons on the right:

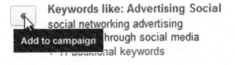

Note   If you're getting confused or need some advice at this point, try calling the Google help line at 1-800-919-9922.

You can use the Automatic Ideas wizard, and you can also type in phrases of your own:

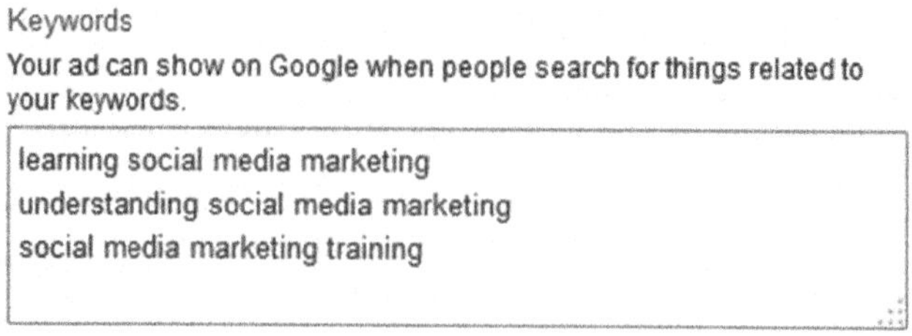

Keywords

Your ad can show on Google when people search for things related to your keywords.

learning social media marketing
understanding social media marketing
social media marketing training

Then click Continue to Ads:

Continue to ads

My advice is, especially if your head is swimming, to make an ad and not to worry about the particulars or keywords too much. This is just for the learning experience. Delete or cancel the ad as soon as are ready and then come back and try again. It's more important to try the full process of making an ad, without worrying about getting it exactly right the first time.

This is the core of making the ad in Google, whether you follow the initial wizard or create a campaign first, and then create an ad group, and then create an ad.

The Destination URL is the link to your e-commerce shopping page. Remember you can always click on the little question mark icons for more information.

Next, enter a headline. Try it; you won't hurt anything. Look at what happens on the right:

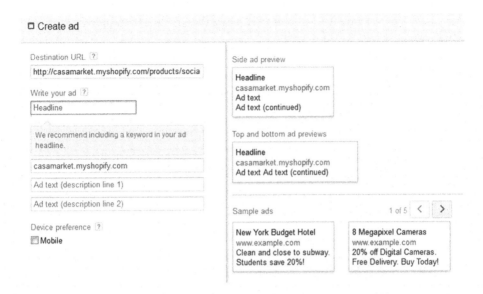

Try experimenting a bit with headlines and with the ad text. Try to think of something that would get someone to click it.

Think creatively. Dream wild! When you are ready, click Save:

Next, you can click Review Campaign:

Review campaign

Here's where you can bid. Just like with eBay, there's an art to bidding, but you can always just use $1.00 to start:

Review

You're almost done! Review the performance estimates and settings for your campaign.

Daily estimates ?	Bid and budget	Campaign Settings
Please specify a bid to get traffic estimates	$ [ ] Bid ? ⚠ Please enter a default ad group bid. Focus on clicks, manual maximum CPC bidding $ [5.00] Daily Budget ?	Canada, United States Google search and search partners English After you finish this campaign, you can change these settings.

Google may try to convince you to spend more and bid more:

Review

You're almost done! Review the performance estimates and settings for your campaign.

Daily estimates ?	Bid and budget	Campaign Settings
0.02 – 0.03 Clicks $0.00 – $0.01 Cost	$ [1.00] Bid ? Focus on clicks, manual maximum CPC bidding $ [5.00] Daily Budget ?	Canada, United States Google search and search partners English After you finish this campaign, you can change these settings.

> Based on historical data, you might not spend your daily budget. Consider adding more keywords or using the following bid:
>
> $7.33 (1 clicks per day, 99% of daily budget)
>
> [Change my bid]

The fact is that this example has its limits. My goal is to provide a cheap/free way to try things out that might result in you actually selling something (woo-hoo!), even if you used this free social media marketing book as an example.

But for better or worse, especially with social media, there's always competition!

There happen to be a lot of trainers, schools, consultancies, etc., that are all interested in people who want to learn more about social media marketing. The prices of their products and services are much higher than a little social media marketing book, so they can afford to bid more on keywords.

When you don't put quotes around a keyword, Google will automatically create variations of it. Google takes something like this:

Learning social media marketing

And makes a variation like this:

Social media marketing

You could end up in competition with someone who is selling social media marketing as a service and charging a lot more money. The bidding will be higher than someone who is selling a service to learn about social media marketing, or selling a book.

This is just a little taste of how eBay—I mean AdWords—works.

The way it relates to analytics is that there's data around the competition for keywords, their average bidding price, and so on. There are analytics in AdWords, in addition to the information that AdWords generates. At the end of the day, it can help your career and business.

You can try putting quotes around words so Google uses those exact phrases only:

Keywords

Your ad can show on Google when people search for things related to your keywords.

"learning social media marketing"
"learning facebook marketing"
"social media marketing training"
"understanding social media marketing"

To play this game a bit better, try a tool called the Keyword Planner, discussed next.

Use Keyword Planner

To try Keyword Planner, choose AdWords ➤ Tools ➤ Keyword Planner.

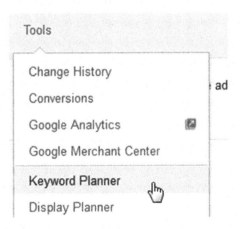

This is the toy you can play with to figure out keywords. I mean, this is the tool you can utilize to attain clarity on keyword potential.

I played around with it and found that the competition is pretty high.

Ad group (by relevance)	Keywords	Avg. monthly searches ?	Competition ?	Suggested bid ?	Ad impr. share ?	Add to plan
Social Media (14)	social media trai...	7,840	Medium	$8.96	0%	»
Social Marketin...	social media ma...	1,120	High	$18.81	0%	»
Networking Busi...	business social ...	1,090	High	$18.40	0%	»
Marketing (7)	online marketing...	1,190	High	$8.04	0%	»
Keywords like: I...	book by isbn, is...	100	High	$1.67	0%	»

I dug a little deeper and found that the average "CPC" (cost per click) of "social media marketing training" was $7.00. So I thought, okay, I'll try an ad where the bid is $7.00 and I'll set the price of the product to $7.00 in Shopify.

	Keyword	Clicks	Impr.	Cost	CTR	Avg. CPC	Avg. Pos.
	"cheap social media marketing book"	0.00	0.00	$0.00	-	-	-
	"i want to learn social media marketing"	0.00	0.00	$0.00	-	-	-
	"learning social media"	0.02	14.81	$0.14	0.2%	$6.27	1.83
	"social media marketing book"	0.00	0.91	$0.00	0.0%	-	1.20
	"social media marketing career"	0.00	0.00	$0.00	-	-	-
	"social media marketing lessons"	0.00	0.00	$0.00	-	-	-
	"social media marketing training"	0.56	15.54	$3.89	3.6%	$6.89	3.20
	"understanding social media marketing"	0.00	0.73	$0.00	0.0%	-	2.75

I adjusted the ad:

Side ad

Learn Social for $7.00!
casamarket.myshopify.com
Social Media Marketing ebook by
five-star author Todd Kelsey, $7.00

Then I created a couple more ads, trying different copy:

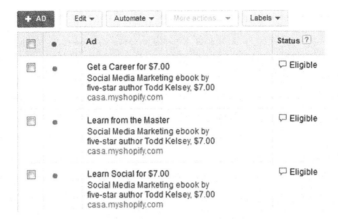

All this is to say that you should experiment, leverage the analytics, try things out, and see what works the best.

AdWords' Tricky Timing Settings

The other important thing to remember when playing with AdWords is to keep track of when a campaign starts and ends. For whatever reason, Google added a couple extra steps for actually ending a campaign. You could say that it is because many campaigns are ongoing. Or you could say that they are doing this because they are apt to make more money.

Either way, I recommend reading this article:

https://support.google.com/adwords/answer/2404203?hl=en

Basically, when you create your campaign, be aware that you will want to go into Campaign Settings and click Edit next to the Campaign Type:

Campaign settings

Campaign name   **Social Media Mktg Primer**   Edit

Type ?   **Search Network only - Standard**   Edit

Then you will want to click the All Features radio button:

Then click Save.

In my opinion All Features should be enabled by default, or there should be an easier, more apparent, way to set the schedule (the start and end dates). But until that happens, be sure to take these steps to make sure Google doesn't drain your bank account every day, if you are just learning.

Next, after enabling this feature, go into Advanced Settings (cough, cough, it should be a basic setting).

Click Schedule:

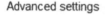

Doh! Google has decided that your campaign of giving them money has no end date:

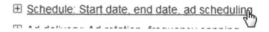

Click the Edit button and set an end date.

Ah! Much better. Just be aware of these hoops you have to jump through.

Advanced settings

⊟ Schedule: Start date, end date, ad scheduling

Start date **Aug 13, 2014**

End date **Aug 18, 2014** Edit

Ad scheduling ⁇ **Showing ads all the time** View ad schedule »

It's Not Quite as Simple as This

I've covered a lot of ground, but I think it's worth trying things out, to get a sense of how tracking conversion works.

And yes, it's not as simple as this. Because clicks don't necessarily mean conversion into sales. Someone might click on the ad and not actually buy something.

The art and science of AdWords involves working with a variety of analytics. In theory, if someone buys something, you can track ad budget against revenue, and that is the primary basis for $40-50 billion of Google's revenue each year, which represents a large amount of revenue made by businesses.

Here was the competition I was up against in my little test:

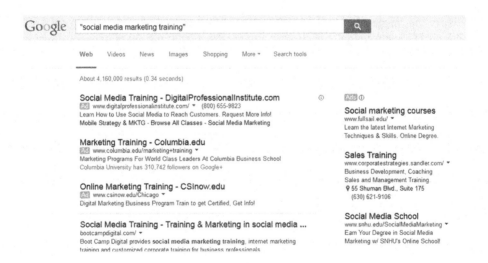

Where oh where is my ad?

Learn from the Master
casa.myshopify.com/ ▾
Social Media Marketing ebook by
five-star author Todd Kelsey, $7.00

Eventually I did find it, on the third page of results. I had a friend click on it to test things out.

Depending on what your goal is—to learn or to actually sell something—you might want to give yourself the freedom of not worrying too much about the keywords and not worrying too much about the bids. Just set things up live and get a friend or two to search on Google for your keyword until they find the ad. Then buy your product (at a low price!), and give yourself 24 hours to look back in AdWords to see the results.

What you're hoping is to see some converted clicks:

0

That is the magic of Shopify and AdWords. It's pretty much at the core of $50 billion of Google revenue each year, and maybe a trillion dollars of e-commerce revenue around the world.

You don't have to be a big company to do it. You, can in fact, as an analyst or online marketer, help people explore this kind of thing, even if you hire an AdWords specialist and just review the information.

In short, it's a technique that millions of businesses—large and small—use, and it's a really solid approach. It can be fun as well.

Learning More

My apologies if your head is swimming. You should check out the following links for more information.

Shopify

- Help center: https://help.shopify.com/
- Connecting to AdWords: http://docs.shopify.com/ manual/your-store/dashboard/google-adwords

- Free AdWords credit: `http://ecommerce.shopify.com/c/ecommerce-marketing/t/aha-this-is-how-to-get-your-google-adwords-and-facebook-credit-121154`

- AdWords: `www.google.com/adwords`

- AdWords Help Center: `https://support.google.com/adwords#topic=3119071`

Conclusion

Congratulations on making it through this wild ride of the last two chapters! Have I convinced you to try things out? Did you have a good experience? Was it intimidating? Inspiring? Best wishes!

Fun with E-Commerce Analytics Part III: Gumroad

This chapter takes another look at e-commerce analytics, courtesy of Gumroad, a free, super easy-to-use platform. In some ways, this chapter is a back up to the previous two chapters, because it provides another way to learn about analytics in a live setting. In the case of Gumroad, there's an easy way to link to Google Analytics, and it also has its own built-in analytics. Even though you can't track conversion directly back to a specific ad, you can get a sense of how Google Analytics works with e-commerce, and it is completely free. And fun!

© Todd Kelsey 2017
T. Kelsey, *Introduction to Google Analytics*, DOI 10.1007/978-1-4842-2829-6_7

Conversion vs. Infersion

One of the advantages of writing a book is that I can get away with making up a word like "infersion". But I think it has merit. I think that conversion-related analytics are among the most important, if not "the" most important to learn, because of how valuable they can be to your bottom line.

On the one hand, you have "direct" conversion tracking analytics available, via platforms like AdWords. See Chapters 5 and 6. Google wrote the book on this, and the power of this direct conversion tracking is that you can trace exactly how effective a specific ad is. Furthermore, you can track how effective individual keywords are. These "conversion analytics" help you optimize a campaign. For example, when you connect these pieces, analytics allows you to optimize, by deleting keywords that don't perform as well and focusing on ones that do better. The analytics help you make an advertising campaign more effective and more efficient, and this is very valuable. It helps you generate more revenue and use the existing budget more wisely. It is the depth of insight that the analytics provide, and this insight helps guide your actions.

On the one hand, if you read these chapters and work through them, you see that it takes more effort to set up and track such things. On the other hand, it's worth it. It's a best practice, and definitely worth learning about and trying.

This is what I call "infersion" tracking. That is, you infer, or guess, what is going on, but you don't really know. In a way, most forms of advertising are based on this principle—billboards, radio, television, etc. You advertise, hope for the best, look at the sales figures, and hope they go up.

In the case of Gumroad, we are talking about infersion. That is, Gumroad has built-in analytics that will tell you how many people visited the site and how many people went on to actually purchase an item. As a learning experience, these basic analytics are a great starting point.

In order to generate some sample data, I made a post on Facebook to friends and asked them to buy the digital ebook (I said I would pay them back on PayPal). I also ran a Facebook ad. I didn't spend a lot of time tweaking the Gumroad site—my focus was on generating the data.

In Gumroad, I got 31 views and one sale during the day or two ran this experiment. I was the only one who purchased the item, but for the purposes of the example, we'll pretend that someone else purchased the item.

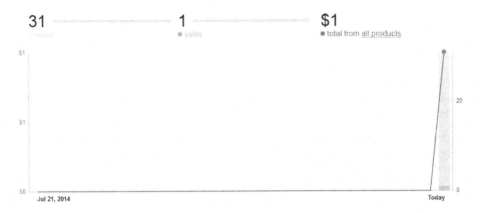

So the conversion rate is 1 sale for every 31 visits. If you divide 1/31, you get .032, which is basically a 3% conversion rate.

The actual conversion rate isn't as important as the fact that it provides you with actionable insight. You can measure the effectiveness of your efforts to improve things, over time. For example, I could spend more time improving the site, adding more visuals and a more compelling value proposition and follow the best practices of digital marketing. I might also make a separate landing page, where I could focus on these things, and then and only then refer people to the "catalog".

After I made these kinds of efforts, I would hope to improve the conversion rate of people coming to the site. Gumroad's built-in analytics provides some limited insight, but it is still actionable.

As for the the Facebook ad I made, Gumroad can't tell me anything about it. It also can't tell me whether one ad was better than another if when have more than one version.

Technically, if I was only running a single advertising campaign, I could "infer" how well the ad campaign was based on looking at sales. Since it is my only ad, I can reasonably guess that it is the force behind sales. But what if there's traffic coming from different sites, or what if a blogger picks up the site? That's when it gets harder to track the impact. So "infersion" has its limits.

Still, I think it is worth looking at.

As you'll see, there are a few things you can do in Google Analytics (which "expands" Gumroad's basic analytic capabilities), and they can provide a bit more insight.

Google Analytics, AdWords, and Gumroad

For those who are interested in looking at the finer points, I mention an area to look into called *e-commerce tracking*. My approach and intent with this book has been to keep things as simple as possible and to focus on learning experiences that can help you understand some of the fundamental concepts in analytics.

Technically speaking, in Chapters 5 and 6, the focus was on AdWords and Shopify. There were analytics "in" AdWords, and in conversion tracking, you can trace this without ever touching Google Analytics. However, when you are running AdWords campaigns and have Google Analytics installed, you can get additional insight by linking AdWords and Google Analytics.

I don't want to spend too much time on this because I think learning by doing is the best way to get a handle on things—but suffice it to say that there's more than one way to do things. I think you can think of it like layers in an onion. You can track conversions with AdWords, but linking to Google Analytics can provide *more* insight. (Do Google searches on "e-commerce tracking in adwords" or "linking AdWords and Analytics" for example.)

The same principle applies to Gumroad—there's a basic way to connect Google Analytics, but there's also a way to set "goals" to provide further insight (see the "Learning More" section at the end of this chapter).

My general recommendation is to Gumroad on its own, then try Google Analytics, then maybe try setting "goals" in Google Analytics. If possible, try selling an actual product, even if it is just friends who buy it (or yourself!).

Let's get started!

Get Started with Gumroad

One of the reasons that Gumroad is simple is because it is built on selling digital downloads.

To get started with Gumroad, go to gumroad.com and click Start Selling:

Then enter an e-mail address and password and click Create Account:

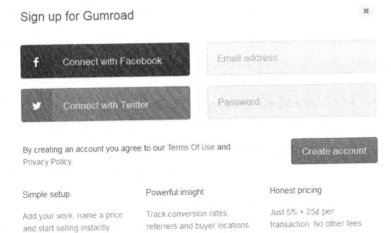

Next, name your store and enter a description in the bio section:

If you are going to do this for real, you will probably want to expand the bio. Go back to Settings in Gumroad and edit/expand the text. Gumroad at present is a bit limited—it's mainly a streamlined shopping cart for digital downloads, so it's not really built to be a "landing page" for an ad campaign, at least in terms of text formatting, having visuals, and fine-tuning your value proposition. But if you do want to add credibility, you can expand the bio section and see what I did for the text that appears at http://gumroad.com/casamarketing.

When you choose a username, this is what will appear in the link:

My store link became gumroad.com/casamarketing.

Next, confirm language settings and set the time zone:

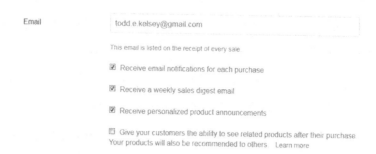

Then enter an e-mail address and choose the desired options:

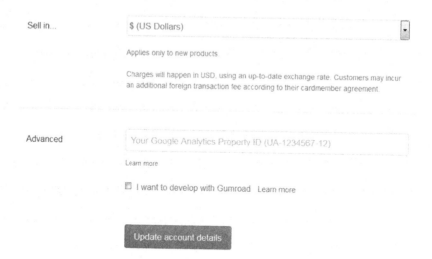

Then, for now, just click Update Account Details:

You'll notice that this section has an area to connect Google Analytics, and the way you get back is to go to Settings in Gumroad. But you need a code from Google Analytics, so we'll discuss that next.

Connect to Google Analytics

You may want to start out by just skipping ahead, adding a product in Gumroad, and buying one yourself to generate data, or post to social, or otherwise just get a few sales in Gumroad, before jumping into Google Analytics. Nothing wrong with that.

If you can, I recommend going the extra mile and putting the foundation in for Google Analytics. In order to do this, you need to know the link for your Gumroad store (choose Gumroad ➤ Settings).

In my case, it was gumroad.com/casamarketing:

Username	casamarketing

View your profile at: gumroad.com/casamarketing

The reason you need it is to tell Google Analytics what site it is tracking.

So, to get a code from Google analytics, go into Google Analytics:

http://www.google.com/analytics/

At the site, click Access Google Analytics:

Access Google Analytics

(This will appear when you are signed in to Gmail/Google and go to the site. If not, you may have to sign in.)

Next, click on the Admin link:

There are multiple ways to add a new site, but this one may be easy enough. You'll have an account, and you'll need to add a new property, which is a new site:

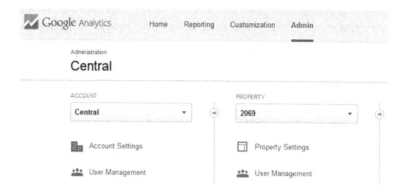

Click on the Property drop-down:

Click Create New Property:

Select Website:

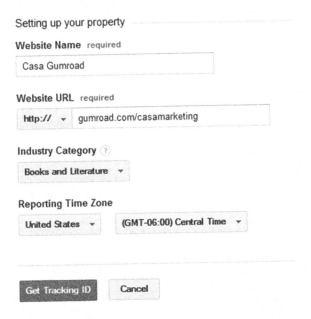

Then set up a name (it's not important as long as it makes sense to you). Paste in the link of your Gumroad site:

Choose a category, set a time zone, and click Get Tracking ID.

Note   Even if you are just running a test, you'll want to set up your own Gumroad store, even if you just upload a "dummy" PDF file for digital download. In other words, use your own link, with your own username. If you try `gumroad.com/casamarketing`, you won't get any data.

Next, you'll get a page with code on it.

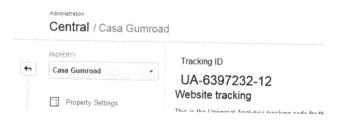

Select the tracking ID:

Copy it into memory. Press Ctrl+C, right-click, then choose Copy in Windows, or press Ctrl+click and choose Copy on a Mac.

Then you can keep it in memory or paste it somewhere.

Back to Gumroad

To paste your Google code, go into Gumroad and then into Settings:

In the Advanced section, paste in the code:

After it is pasted in, click Update Account Details:

Advanced

UA-6397232-12

Learn more

☐ I want to develop with Gumroad   Learn more

Update account details

Add a Product to Gumroad

To test your Gumroad account further, you need to add a product. To get started, click Products:

Click Add a Product:

Then click Product type on the left:

It's very easy, relatively speaking, to add products in Gumroad, compared to most e-commerce platforms.

Just type in a name for the product and add a file to it.

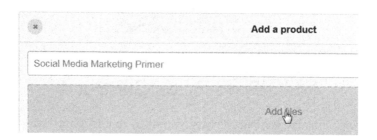

It will give you progress on the upload:

You then choose a price and click Add:

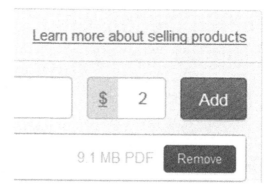

You might just want to make the price $1.00 to keep things as cheap as possible.

Next, you need to upload a cover for the product. Gumroad likes square images. Feel free to use the "cover 250x250" image, review the chapter on "Content" in my *Introduction to Social Media* book, use an image created in a program like SnagIt or Gimp, or try picresize.com.

Click Upload a Cover:

I also recommend clicking on the Describe this Product field and entering a description.

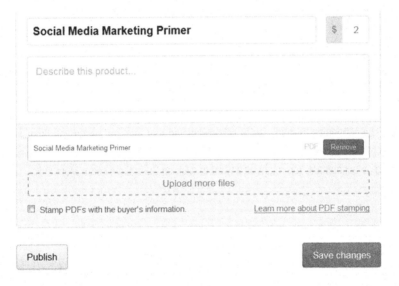

When you're ready, click the Save Changes button:

Then click Publish:

Setting Up Payment

This may not be strictly necessary for testing, but you'll probably want to set up payment. Depending on where you're at in the process, you might need to click on Products in Gumroad:

The easiest approach is to set up direct deposit. You just enter a checking account and routing number, which you can get from a check:

Then you click Change on Tax Settings if you want to. If you're just testing your account, you can ignore the tax settings for the moment.

■ **Note**   By continuing to read, you're acknowledging that I am not giving you financial advice about e-commerce taxes. If you plan on selling anything on an ongoing basis, I recommend consulting an accountant. You might also want to click on Gumroad's Learn More link.

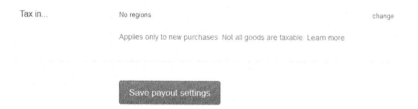

When you're ready to move forward, click Save Payout Settings.

Check Your Gumroad Site

At some point, if you haven't done so already, you'll want to check your Gumroad site to see what it looks like.

If you're not clear yet on the link, go into Gumroad and click on Settings:

Then scroll down and see what it says under username:

Username	casamarketing

View your profile at: gumroad.com/casamarketing

Then you can copy and paste that link into a browser. I recommend testing with multiple browsers, specifically Firefox (`firefox.com`) and Chrome (`google.com/chrome`). Work with Gumroad in one, and then "test" the site in the other. The reason is that when you're logged into Gumroad, the site appears slightly differently than when you're not, so by going into a separate browser where you're not logged in, you get a better picture of the site.

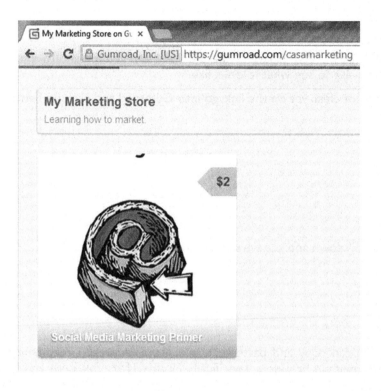

Get Some Traffic to Sell

As with earlier chapters and based on techniques in my *Introduction to Social Media* book, I suggest trying to get a bit of traffic to the site, through a combination of making an ad on Facebook, or using AdWords, or both, as well as making an appeal on social media to your friends.

Gumroad does have some built-in connections to make it slightly easier. Try going to Timeline:

Edit   Options   Timeline

Timeline

Then click Share on Facebook:

You can also select and copy the short link (gum.co/etc.) to text, e-mail, or post to friends.

Then make some kind of kindly appeal for people to click and buy the product, and click Share:

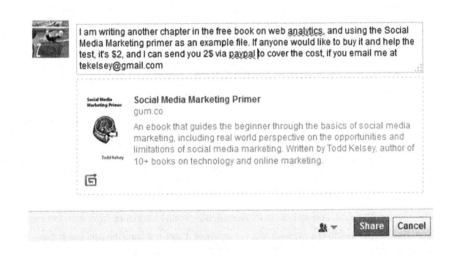

Reviewing the Analytics

After you get some traffic and at least one purchase, you can go to Analytics in Gumroad:

Be aware that you do need one purchase to "activate" analytics in Gumroad. Divide the number of sales by the number of views to get your conversion rate.

Then, go back to the drawing board, improve the site, lower the price, do some research on making a landing page, and then link to your Gumroad site again. See how the statistics change.

For example, you can try a search on landing page tools:

https://www.google.com/#q=landing+page+tools

Or try Googling "landing page best practices" as well.

Link to your Gumroad store and see if you can improve your conversion rate.

Reviewing Google Analytics

■ **Note**   Google Analytics, and web tools in general, will change interfaces from time to time, so you might have to dig to find certain features. For example, there are some helpful links at the end of this chapter that Gumroad provides about connecting Google Analytics to Gumroad. The exact position of the information may change, though. I still recommend the article, but keep in mind you might have to dig around for it. The moral of the story is, play around and search for things, and when all else fails, Google it, because someone else probably had the same question and blogged about it. Remember, you can blog about it too!

Take screenshots! Share your thoughts on what you learned! Make a blog! Share it with others! This is how many blog entries came into being that so many people have found helpful. Someone had to do it, even if they didn't feel like an expert. Try it! Build your portfolio!

Go to Google Analytics and select All Website Data in the property you want to look at:

Next, go down to Behavior on the left, then to Events, and finally to Overview:

You'll see something like what you see in Gumroad:

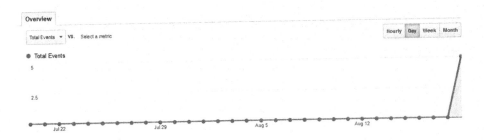

The difference is that you can dig a little deeper. For example, try scrolling down and clicking on Event Action:

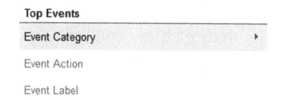

At this level, you'll get a sense across all products of how many times people clicked "I want this" in Gumroad, and of those people, who actually made a purchase:

Event Action	Total Events	% Total Events
1. iwantthis	3	60.00%
2. purchased	2	40.00%

This is an example of analytics related to *abandonment*. To learn more, Google "cart abandonment" and "abandon rate e-commerce". It's something that e-commerce businesses pay close attention to, in terms of looking at how many people go along the process of visiting the site, and where they tend to drop off.

For example, if people click on the Buy button but don't check out, there could be a misunderstanding about how to check out, or an issue with shipping prices, etc. By looking at where a person leaves the site, you can pinpoint actions you can take to reduce abandonment.

The information from Gumroad ➤ Google Analytics is pretty limited, relatively speaking, but it gives you a general idea. In more extensive analytics environments, such as Shopify, you may be able to get more information on the paths people take, as well as get a sense of where the highest percentage of people are dropping off.

In our example, you could also go to the same spot in Google Analytics and click on Event Category, and then click on an individual product: (such as product-bVxL)

Top Events	Event Category	Total Events	% Total Events
Event Category ▸	1. product-bVxL	4	80.00%
Event Action	2. product-undefined	1	20.00%
Event Label			view full report

This will give you a closer view of an individual product.

Then you can click on Event Action from there.

It will give you similar information for a specific product. This is important, because if you have multiple products, it's important to track the difference between them. In some cases with analytics, you'll want to drill down to a specific product level.

Learning More

To learn more about Gumroad, see `http://help.gumroad.com/`.

There's also some information from Gumroad on analytics:

- Setting up "goals" in Google Analytics (I highly recommend exploring and trying this): `http://blog.gumroad.com/post/87921227603/how-to-set-up-goals-in-google-analytics`
- General analytics: `http://help.gumroad.com/using-google-analytics-and-gumroad`

Remember that if you follow the description on where to find Google Analytics, it might be out of date. But it's still worth reading.

This link is helpful when working with products:

`http://help.gumroad.com/11162-getting-started/adding-a-product`

Conclusion

Congratulations on taking another tour through analytics and e-commerce!

I encourage you to continue exploring various connections you can make, between shopping carts, analytics and ad campaigns. This is the "core" of web analytics, and Google Analytics can play a central role in helping to make sense of things. If you haven't already, you may also want to *collaborate*— concentrate on what you do the best and work with others to do the rest! For example, if you want to focus on the analytics, find a friend or local business or organization that has something to sell, and work with them on getting a site up and going. Integrate Google Analytics and get a sense of visitors. Then try selling something or get someone else to run the ad campaign and look at conversion tracking. Then optimize the site.

Start out simple, have some fun, and before you know it, you will be considered a web analytics guru—even if you don't feel like one!

In the next chapter, we look at the ultimate goal of this book, or one of the ultimate goals. We explore and inspire you to dig into Google's free learning material, and work toward getting certified in Google Analytics. This book was mainly written to introduce you to analytics, and to help you try a few things, have some fun, and get some momentum. Learning more about Google Analytics is the next step. It will give you confidence, it will help your career, and it might even help you get a new client or a new job for that matter. It's doable, and I highly recommend it!

Exploring Google Analytics Certification

This chapter explores some of the opportunities that Google offers, in terms of learning material, and the Google Individual Qualification (IQ), which can be a nice thing to have on your web site or resume.

Maybe you can take me on faith that certification/qualification is doable. It's especially doable if you have resources to rely on, including people. The benefits of going through the process are many, including a definite positive impact on your career.

Imagine adding a Google certification to your LinkedIn profile:

 Certifications

Google Analytics Individual Qualification (IQ)
Google, License 00525913

Google Adwords Search Advertising Certification
Google

© Todd Kelsey 2017
T. Kelsey, *Introduction to Google Analytics*, DOI 10.1007/978-1-4842-2829-6_8

These qualifications give your resume extra impact, whether you are job searching, interacting with other professionals, or just boosting your credentials at your current job.

It can also be a good thing for your business, when you list your staff on your web site and show that they have these certifications.

Going through the certification process also helps to round out your knowledge. It's smart to be acquainted with new features that you might end up using at some point.

Exploring Google Analytics Qualification

What's the difference between certification and qualification? It's basically interchangeable, as far as I can tell.

To explore Google Analytics Qualification, check out this site: `www.google.com/intl/en/analytics/learn/`.

Analytics Training and Support

These self-service options are available for users of Analytics Standard (the free version).

If you require support via email or phone, consider upgrading to Analytics 360.

Training

 Analytics Academy

Analytics Academy ⬀ offers free, online courses on Analytics and other data analysis tools. You can use Analytics Academy to prepare for the Analytics Individual Qualification (IQ) exam—an industry recognized qualification.

 Analytics Demo Account

The Analytics demo account ⬀ is a fully functional Analytics account that any Google user can access. It's a great way to look at real business data and experiment with Analytics features.

Support

 Analytics Help Center

The Analytics Help Center (the current page) documents all aspects of using Analytics including how to get started, best practices for analysis, and troubleshooting.

 In-product Analytics Help

Search

The Help Center Search in Analytics lets you look up support information from within the user interface. To access Help Search, click **Settings** ⚙ in the upper-right corner and select Help.

I think it's good to go through the process of starting your own account, but you can try the demo account as well.

Be sure to look at the other articles.

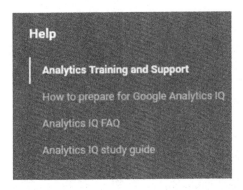

The Analytics Academy is also worth paying attention to. The point is that it's free training given by a global company, and it's in their best interest for you to succeed.

Training

 Analytics Academy

Analytics Academy ☑ offers free, online courses on Analytics and other data analysis tools. You can use Analytics Academy to prepare for the Analytics Individual Qualification (IQ) exam—an industry recognized qualification.

I recommend going directly to this Google site to become familiar with some of the self-study material they have:

https://analyticsacademy.withgoogle.com

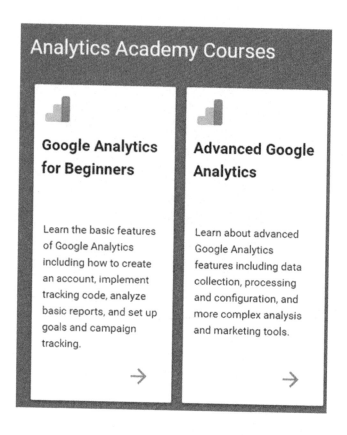

Start Easy

To get rolling, before you take a course, you might want to watch some videos:

https://www.youtube.com/user/googleanalytics

Click the Playlists link on the YouTube channel to see what's available:

https://www.youtube.com/playlist?list=PLI5YfMzCfRtZ8eV576YoY3v
IYrHjyVm_e

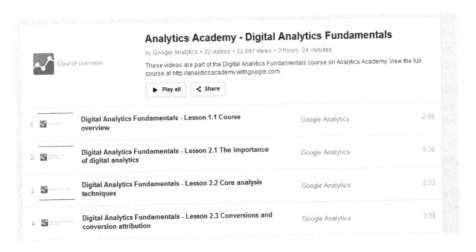

Roadmap for Certification

My general suggested roadmap is to take a peek at some of the resources mentioned in this chapter and then do two things in parallel:

- *Use Google Analytics:* You'll get more out of the materials as you prepare to take the Google Analytics test if you're actually using the program.

- *Go to the Analytics Academy link:* Review the roadmap they set up for studying, and if you like, start by watching some of their videos.

And at some point, sooner or later, you'll want to read, "How to Prepare for the Google IQ" at `https://support.google.com/analytics/answer/3424288`. Remember to review the "Additional Resources" section as well.

Get Ready, Go!

When you're ready, go ahead and register at `google.com/partners`. In my classes I teach, I recommend people take the exam once just to see what it's like. So go ahead!

Then study some more so you can pass it. You can do it! It's doable. I've had many students pass it. If they can do it, you can do it. They all feel good when they pass, and they can put it on their resume and on LinkedIn. If you have coworkers or a manager, it's a great thing to tell them too. If you're in the job market, I know for a fact that the certification impresses recruiters.

Not only do I think the IQ is doable, but I think it can have a good impact on your career, as well as on your business or organization. Even if you're not in the job market yet, the fact that it's a top skill means learning it is important.

As I mentioned in the introduction, LinkedIn shows digital/online marketing as a top skill that gets people hired year after year, and web analytics is one of the core digital marketing skills. Being able to understand the performance of web sites and ad campaigns is important. Analytics is considered a part of business intelligence, which also figured prominently on the list:

- (6) User Interface Design
- → (7) Digital and Online Marketing
- (8) Recruiting
- (9) Business Development/Relationship Management
- (10) Retail Payment and Information Systems
- → (11) Business Intelligence
- (12) Data Engineering and Data Warehousing

Demand will fluctuate over time, but we are talking about the top skills *in any field* that get people hired.

- 2014: https://blog.linkedin.com/2014/12/17/the-25-hottest-skills-that-got-people-hired-in-2014

- 2015: https://blog.linkedin.com/2016/01/12/the-25-skills-that-can-get-you-hired-in-2016

- 2016: https://blog.linkedin.com/2016/10/20/top-skills-2016-week-of-learning-linkedin

Don't forget how nice it will look to have this on your LinkedIn profile. Thousands of people have done it. You can too!

Long-Term Goal: Certified Partner

If you work for or will likely end up at a "services" company, where you help others with marketing, you might want to consider the Certified Partner program. It's a bit involved, but in order to compete, to deliver needed value to other businesses and organizations, you should seriously consider going down this road:

https://www.google.com/analytics/partners/listing/service

Google Analytics Certified Partner

The Google Analytics Certified Partner program is a company level accreditation for business consultants who package, sell and deliver analytics, website testing and conversion optimization services to online businesses. If this is what your company does, consider becoming a Google Analytics Certified Partner to grow your business and boost your clients' website conversion rates and profitability.

Based on what I've seen, analytics is one of those areas that is very, very solid. You're providing crucial insight to businesses and organizations, and in a competitive market, it's a skill set and a service that is in high demand.

It takes some effort and it may take a while to grow things. I just want to encourage you to consider it. Let's say you find yourself up against 10 other social media businesses or 10 other marketing agencies. How many of them have people who are certified in Google Analytics? How many of them are actually certified on the business level? If you want to have a steady, long-term stream of business, to supplement other things, I have a pretty strong feeling that heading in this direction can only help you.

But don't take my word for it—do some background research.

A business that has Google-certified consultants inspires confidence:

Google Analytics Consulting Services

As Certified Google Analytics Consultants, we offer comprehensive analytics consulting in the areas of strategy, implementation, optimization and training for both Google Analytics Standard and Google Analytics Premium.

Google Analytics Standard Our analytics experts can help you gain the insights that matter from this robust measurement tool.

Google Analytics Premium Enterprise Analytics with implementation, support and comprehensive consulting from Blast included.

Learning More

Here are some additional discussion and tips that I think are worth reviewing.

- Benefits of certification: `http://smallbusiness.chron.com/benefits-google-analytics-certificate-56029.html`

- Tips for passing the exam: `http://www.webucator.com/blog/2011/03/tips-on-passing-google-analytics-individual-qualification-exam/`

Conclusion

Congratulations on making it through this chapter! And the book for that matter! If you want a roadmap for doing the Google IQ, timeline-wise, I suggest reviewing the "units" at a rate of one per week, so that you're making steady progress but also give yourself time to use the program.

After reading the chapters in this book, you should be able to generate some traffic, even if you aren't connected to anyone else. I suggest checking to see if there are any local businesses or organizations who might like some help. Go to your local chamber of commerce or an organization that local businesses belong to. Chances are there will be someone who has a web site and would be interested in having someone look at their web analytics free of charge. If there's any way you can set that in motion, I think that actively looking at some kind of traffic will help give you perspective.

But don't wait—start reviewing the material.

Best wishes!

■ **Special Request** Thank you for reading this book. If you purchased this book online, please consider going on where you purchased it and leaving a review. Thanks!

I

Index

A

Adobe Analytics/Omniture, 6–7

AdWords, 84, 110
  creating account/getting tracking
    code, 85–87
  creating campaign, 92–99
  and Google Analytics, 84
  revenue, 104
  to Shopify, 88
    to modify or not to modify, 89–91
    Order Confirmation page, 91
  tracking conversion, 103
  tricky timing settings, 101–102
  using Keyword Planner tool, 99–101

Analytics, 1–3
  Academy, 131
  free and corporate tools, 6
    Adobe Analytics/Omniture, 6–7
    Google Analytics, 6
    open source analytics, 7
    social analytics, 7–8

B

Blogalytics
  blog, starting, 9–10, 12
  Google account/Gmail address,
    creating, 9
  Google Analytics account, starting, 13–17
  Google Analytics to blogger, 17–18

Bots, 44

Bounce rate, 40

Business intelligence, 5–6

C

Campaign analytics,
    reviewing, 37–38

Conversion tracking, 54

Conversion *vs.* infersion, 108–109

D

Digital/online marketing, 4

E, F

E-commerce analytics
  AdWords, 84
    creating account/getting tracking
      code, 85–87
    creating campaign, 92–99
    revenue, 104
    to Shopify, 88–91
    tracking conversion, 103
    tricky timing settings, 101–102
    using Keyword Planner
      tool, 99–101
  conversion *vs.* infersion, 108–109
  conversion tracking, 54
  e-commerce tracking, 110
  Shopify site
    buying own product, 80–81
    customize navigation, 61–65
    to get started with, 56–61
    live and purpose, 54–55
    tweaking and payments, 65–74,
      76–79

Event Category, 126

© Todd Kelsey 2017
T. Kelsey, *Introduction to Google Analytics*, DOI 10.1007/978-1-4842-2829-6

Get the eBook for only $5!

Why limit yourself?

With most of our titles available in both PDF and ePUB format, you can access your content wherever and however you wish—on your PC, phone, tablet, or reader.

Since you've purchased this print book, we are happy to offer you the eBook for just $5.

To learn more, go to http://www.apress.com/companion or contact support@apress.com.

Apress®

CPSIA information can be obtained
at www.ICGtesting.com
Printed in the USA
LVOW13s1839210318
570660LV00004B/172/P